ROUTLEDGE LIBRARY
SOCIAL AND CULTURAL

Volume 10

THE CHANGING NATURE
OF GEOGRAPHY

THE CHANGING NATURE
OF GEOGRAPHY

ROGER MINSHULL

Routledge
Taylor & Francis Group

LONDON AND NEW YORK

First published in 1970

This edition first published in 2014
by Routledge
2 Park Square, Milton Park, Abingdon, Oxfordshire OX14 4RN

and by Routledge
711 Third Avenue, New York, NY 10017

Routledge is an imprint of the Taylor and Francis Group, an informa business

First issued in paperback 2015

British Library Cataloguing in Publication Data
A catalogue record for this book is available from the British Library

ISBN 978-0-415-83447-6 (Set)
eISBN 978-1-315-84860-0 (Set)
ISBN 978-0-415-73356-4 (hbk) (Volume 10)
ISBN 978-1-138-98887-3 (pbk) (Volume 10)
ISBN 978-1-315-84826-6 (ebk) (Volume 10)

Publisher's Note
The publisher has gone to great lengths to ensure the quality of this reprint but points out that some imperfections in the original copies may be apparent.

Disclaimer
The publisher has made every effort to trace copyright holders and would welcome correspondence from those they have been unable to trace.

THE CHANGING
NATURE OF
GEOGRAPHY

Roger Minshull

HUTCHINSON UNIVERSITY LIBRARY
LONDON

HUTCHINSON & CO (*Publishers*) LTD
3 Fitzroy Square, London W1

London Melbourne Sydney Auckland
Wellington Johannesburg Cape Town
and agencies throughout the world

First published 1970
Reprinted 1972

Printed in Great Britain by litho on smooth wove paper
by Anchor Press, and bound by Wm. Brendon,
both of Tiptree, Essex
ISBN 0 09 102710 1 (cased)
0 09 102711 X (paper)

CONTENTS

PREFACE

The last sentence of *The Spirit and Purpose of Geography,* first published in this series in 1951, states that the geographer's work 'subjects him to a discipline and yields him a philosophy'. This sentence made such an impression on me that since reading it one of my main preoccupations has been to study the nature of the discipline and to work towards some satisfying philosophy.

In the last twenty years such rapid changes have taken place in the methods and aims of geographers that teachers, students and laymen alike, for whom this series is intended, feel the need for a review of the situation. However, a statement and examination of changes and trends could be very insubstantial for the reader with little grounding in the theory of the nature of geography. This book, therefore, is also intended as an introduction to the nature of geography. A selected bibliography is provided for those who would like to pursue the matter further.

Many geography students really face the question of the nature of their chosen discipline for the first time when they start their special study, dissertation, or whatever it may be called. Their main danger is of ending up with a woolly mixture of geography, history and farming and industrial technology. For this reason I have tried to be as precise as possible about the nature of this discipline which we call geography, sometimes to the point where some geographers would say that the definitions are too exacting and exclusive. But I insist that geography is an exacting discipline and not just a polite name for general knowledge.

In this connection two words have been given precise meanings in this book. The word 'geographer' means a professional earning his living by doing research, and either writing or lecturing about his findings. Teachers and students of geography are referred to as such. The word 'space', when not qualified, refers to an abstract idea of space, and not to a particular location, and certainly not to the area between the planets.

While some geographers feel a need for geography to be of economic value in planning and development, as a teacher I feel the need for geography to be relevant to much wider issues. The findings of the geographer must be important to any philosophy of the nature and condition of man. The questions posed at the end of this book, then, are concerned not so much with immediate practical problems, as with the need to understand the nature of man's existence in the universe.

As usual I am indebted to Mrs M. M. Tyson for her typing, and to Professor W. G. East for encouraging me to continue writing, and for his generous spirit in accepting this book for the series.

ROGER M. MINSHULL

I

INTRODUCTION

Many geographers believe that the present rapid change in geography, the quantitative revolution, is resulting in genuine progress. Changes in the past may have been less revolutionary, but it is certainly very difficult to say that progress was made each time. T. W. Freeman goes as far as saying that trends in geography have been like swings of the pendulum and that most of the things we hail as new have been tried or suggested in the past.[1]

There have been changes in content, approach and techniques. Changes in technique are easily explained as the result of a desire to use the best tools available at any one time, and must occur as technology improves. But changes in content, such as the inclusion or exclusion of human geography, or in approach, such as the swing from study of landscapes to the study of spatial relationships, are much more disturbing. If all three elements of content, method and technique are in constant change, then how can we pretend that a discipline called geography exists?

Recently, in *Frontiers of Geographical Teaching,* Wrigley[2] has suggested that an overall definition of geography is not necessary, partly because it leads to rigidity and stultifies growth. Wrigley may be the spokesman for many other geographers who think this point so obvious that there is no need to write it down. However, the majority of articles on the nature of geography are striving to emphasise guides which will keep geographers on the correct course during this constant change. The main problem is to define the nature of geography in such a way that the definition is neither

[1]Superior figures refer to notes at the end of the chapters.

too narrow so that it applies only to a few specialisms, nor so wide that in practice it is useless.

Another problem is the need to formulate a definition which is valuable to the layman, the professional geographer, and to all the students in between. For the interested layman, any short, neat definition here would then need a book of qualification and explanation. Conversely, students can have been studying geography successfully for several years without necessarily comprehending the whole of geography, or being able to give a concise definition. Added to this, we have the increasingly difficult situation of a growing number of specialist professional geographers each defining the subject largely in terms of his own interests.

Any specialist, or any author, must try to avoid defining geography in terms just of what he likes to do, or what he thinks it is, without careful, objective thought. But it must be stated here that bias of this kind can creep in. Avoiding this approach then, basically there are two ways in which to attempt to define geography. The first is to study what geographers have done in the past.[3] There are several dangers in this approach, however. One possibility is that past geographers have been selective in their work and have largely ignored certain topics, say, social phenomena. But the main error we can fall into here is that in trying to define 'geography' we take the word 'geographer' for granted and thus beg the question all together. Logically, it is impossible to call any worker, past or present, a geographer, until we can say what geography is, and then decide whether he has been doing it or not.

The second approach is to ignore all the work which has been done, for the moment, and to work out in principle what geography *ought* to be.[4] One might expect this approach to give a more complete definition, in that thinking out what ought to be studied may indicate certain areas ignored until now. Several people have tried to proceed in this way, and some details will be given later, but there are as many difficulties with this deductive method as with the inductive method. One can start with the word geography; or with the world as an object of study; or with the concept of studying phenomena in space, and so on; with the possibility of such different starting points leading to increasingly divergent conclusions.

The possibility of some working definition, which may be improved as more 'geographic' work is available for inductive reasoning, or an agreed starting point is established for deduction, lies in approaching the problem from both directions, with the hope of finding some common elements for our guidance. The task is most involved, for just as three geographic texts on, say, landforms, economic activities, and the regions of a continent differ so widely, so three papers on the nature of geography in recent editions of the journals seem to be so different that contradictions are inevitable when all the ideas are brought together. For example, F. Lukermann[5] considers that neither content nor method are important, and that geography is defined by the questions it asks. Wrigley,[6] in contrast, writes of 'the welter of observational material with which geographers commonly deal'.

Wrigley's article is very useful in showing how change can be dangerous and misleading. The 'welter of material', or the subject-matter of geography, is constantly changing, especially the subject-matter of economic and social geography as farming, industry, transport, towns and cities change. Now the geographer who defines his work by this subject-matter, the content, is in danger of changing the whole approach and purpose of his work as the purposes and problems of the people he is studying change. This confusion is made worse by the use of the word geography to mean not only the discipline, but the objects of study as well. This can be seen most clearly in *Problems and Trends in American Geography,* edited by S. B. Cohen.[7] The title suggests that the book is about problems and trends within the academic discipline. In fact the majority of the chapters deal with such problems and trends as rural depopulation, urban growth, conservation and so on; that is, with the problems and trends within the subject-matter itself. Taken to the extreme, definition of geography by the subject-matter alone could lead to complete reverses in the purposes of study as trends in the field are reversed.

In the concluding article in the same book, Haggett[8] echoes Wrigley in emphasising that geography should not lose sight of its basic purpose by being misled by short-term, superficial changes, whether these are trends to planning, changes in teaching or changes in what man is doing on the earth's surface. J. D. Chapman[9] suggests a useful analogy here. He likens geographers to people setting out in sailing ships. Some of them, taking advantage of every wind, can travel very quickly indeed, but as the winds change they zigzag all over the ocean, and don't get anywhere in particular. This is speed and change, but it isn't progress. Others have a destination, and this means sailing against a strong wind at times. Their average speed may be very slow, but they have a destination, they have a purpose, and in getting toward it they are making progress.

A common complaint about the work of sixth-formers and college students is that they do not treat their subject-matter 'geographically'. This is partly the fault of their teachers and tutors, but partly the result of the students being unfamiliar with the philosophy of the subject. At different levels the school pupil faced with 'give a geographic account of . . .',[10] the student writing a special study, and the graduate doing research, all have the same problem of using facts and information as geographers, in contrast, say, to historians or natural scientists.

Thus we begin to see that an understanding of the nature of geography is needed for several reasons. Firstly, to keep pupils, students and even research workers on the right track *if* they want to be geographers. It is very easy indeed for what was intended to be geography to end up as quite sound history or an accurate account of economic processes. Haggett[11] urges that we should avoid so-called geography, which in fact is 'genuine historical research'.

Secondly, to provide the basic framework into which all our acquired knowledge about the external world can be fitted. At least since the time of Kant it has been explicit both that our direct knowledge of the world is limited, so that we need information from others, and that our information comes in a haphazard fashion, not necessarily in a logically satisfying order. Therefore some logical framework is necessary early in life so that the pieces of the jigsaw can be put together properly. There is the practical difficulty of giving this philosophical framework to young children. This makes a well-planned curriculum vital, but emphasises that the student should be introduced to the philosophy of his particular specialism as soon as possible, even though this necessitates elaboration later on.

A third reason is to help teachers, in their turn, to understand what they are teaching, to understand the value of geography, and its relevance both to education and to everyday life. During this particular period of change, which is accompanied by a rapid increase in the number of professional geographers, specialist teachers, and diverse branches of the subject, there is a need to emphasise the common subject-matter, methods, and, above all, objectives of geographers. Finally, Haggett would add two points.[12] First the need to attract those who will eventually become research students. Second, to have such a clear definition that vaguely geographic work can not be made an *ad hoc* vehicle for such things as international goodwill, regional planning, bridging the two cultures gap and above all 'short-term educational profit in school'.

Stating the obvious may help to clarify a point here, and indicate whether to start with a definition of geography or an examination of geographers. People who call themselves geographers may not necessarily be doing what is *a priori* defined as geography. For example, most methodologists insist that geography must include a study of distributions and spatial relationships, yet until very recently so-called physical geographers studied such things as the development of landforms, and today would more strictly and properly be called geomorphologists. Therefore it must be stressed that an individual may start off as a geographer, but as his interests change, and the type of work he produces changes, this does not necessarily mean that geography has changed. More likely it means that the individual has become an historian, a climatologist, an economist, a demographer or a mathematician.

From this, it is suggested that there can be three distinct motives for any type of study. First, there is the interest in the content of study. Ignoring the truism, for the moment, that very rarely is the content unique to one discipline, we know that some students approach history through their interest in people and past events; some approach natural sciences through interest in machines and plants; others approach geography through interest in other parts of the world.

Second, there is the delight in methods and techniques applied more in one subject than others. These may be the scientific method and experimental techniques most obvious in physics and chemistry, the historical method, mathematical techniques, or the use of maps and photographs in geography.

Third, there is the desire to work toward some greater knowledge, some genuine understanding of the world. The first two approaches can, of course, lead to this third motive, and one sincerely hopes that they do. But it is quite clear that many people are satisfied simply by acquiring factual knowledge from whatever they study, and certainly one sees many people so delighted with new techniques, machines and equipment that they forget entirely that methods and techniques are simply means to the end of collecting information, sorting it out, and trying to understand the world a little more in the process. W. M. Davis,[13] many years ago, was concerned about the people who took more interest in methods of mapping, rather than in the things mapped. We have a new variation on this theme with those whose interest lies in how to compute quantitative data rather than in what the processed data reveal. The person who stops when the calculation is complete is a statistician; the person who uses the calculation to help in his understanding of the earth's surface is on the way to becoming a geographer.

As people have approached geography, and have become geographers, through interest in the three elements of content, methods and purpose, these will be examined separately in this book. Content and methods of what is accepted as geography have changed and are still changing. While trying not to prejudice succeeding chapters, it is the author's contention that although content can change within limits, and methods will change, we must hope to find or define an underlying purpose common to all geographers. The discipline must be defined by something which is both unique to geography, and constant in time.

1. Freeman, T. W., *A Hundred Years of Geography*, Duckworth, 1961, p. 16
2. Wrigley, E. A., in Chorley, R. J., and Haggett, P., *Frontiers in Geographical Teaching*, Methuen, 1967, p. 15
3. Hartshorne, R., 'The Nature of Geography', *Annals of the Association of American Geographers*, Lancaster, Pennsylvania, 1939
4. Bunge, W., *Theoretical Geography*, Lund, 1966
5. Lukermann, F., 'Geography as a formal intellectual discipline, and the way in which it contributes to human knowledge', *The Canadian Geographer*, vol. VIII, no. 4, 1964, p. 167
6. Wrigley, op. cit., p. 3
7. Cohen, S. B., *Problems and Trends in American Geography*, Basic Books, 1967
8. Haggett, P., in Chorley and Haggett, op. cit., p. 375
9. Chapman, J. D., 'The Status of Geography'. *The Canadian Geographer*, vol. X, no. 3, 1966, p. 133
10. CSE geography papers, East Midlands Board, 1969
11. Haggett, op. cit., p. 375
12. Haggett, op. cit., p. 375
13. Davis, W. M., *Geographical Essays*, Dover and Constable. 1954

2

SOME CONSIDERATION OF CONTENT

When one begins to consider this matter several familiar attitudes
come to mind. Many historians, particularly those interested in
social and economic history, and whose school geography was
old-fashioned, are convinced that geographers simply duplicate
their work. The fact that so many well-educated people in responsible positions still believe that geographers just borrow bits and
pieces from other disciplines further convinces the author that a
book of this kind is necessary. More definitely, some geographers,
when discussing methodology, state at once that the subject-matter
is shared with other disciplines but is treated in a different way for
the geographer's own purpose.[1]

At the other extreme, there are the laymen who studied
geography only at school, who are quite clear that geography
alone studies places, other parts of the world, the landscape and
what people do—to use simple, general terms. Moreover, it is
suggested (whatever the methodologists may achieve in the future)
that the majority of people are attracted to geography by its
subject-matter rather than by its approach or specialised techniques. Even Lukermann seems to contradict himself when he
twice mentions regions as geography's unique objects of study
and then states 'the study of place is the subject-matter of geography because consciousness of place is an immediately apparent
part of reality, not a sophisticated thesis'.[2] Preston James makes no
contradictions at all when he writes 'geography remains that field
of study that focuses its attention on particular places on the
earth's surface'.[3] The bias to regions will not be examined here,

because the only point to be made is that some authorities are sure that certain content matter is unique to geography. Obviously, with such extreme attitudes, one must examine the problem more carefully.

Throughout European and American literature there is general agreement on the content of physical geography. The five major topics are rock type and arrangement, relief and drainage, climate, soils and wild vegetation. These cover all aspects of the physical environment, the four 'spheres' beloved by certain French geographers.[4] Physical geography can be wide enough to include a comprehensive study of these five major topics, and some works are lengthy treatises on natural science; but in geography man narrows the field. Even in a work confined to physical geography the possible scope is reduced to finite terms by concentrating on the aspects of these topics which are significant to human life. Thus geography is concerned with rocks at and near the surface, and so borrows from geology, but is not concerned with the deep interior, and so has much less to do with geophysics. Similarly it studies the atmosphere, but not the stratosphere, and the interests of the geographer and the meteorologist diverge widely. One result of this attitude is so obvious that it goes largely unnoticed; the physical geographer so instinctively studies the five topics with their importance to human geography somewhere at the back of his mind that often he studies only the continents. He studies only about one-third of the earth's surface because this is where the rocks are known, and are of most importance to man: here is the relief, here the soils and the vegetation. Moreover, while atlases show pressure and winds over the oceans, how many show rainfall there too? It rains at sea, but man has kept few records there, and at sea the rainfall is not important. Kendrew, very precisely, called his book *The Climates of the Continents.*

Oceanography exists as a specialist study but aspects of this are seldom included in physical geography. Again the criterion is whether the topic is important to the daily life of man or to some other topic of geography. Thus the effects of waves on coasts, of warm and cold currents on marginal climates, and of many aspects of the sea on fishing *are* studied, but that is all. The physical geographer studies some aspects of the rock, relief, climate, soil and vegetation, but not all. His selection is made, very often, from his view of the physical world as the home of man. The study can go beyond this, and the French concept of this zone of contact between earth, air and water, the few feet in which plants, animals and men live, is a most valuable guide.

While there is general agreement on the content of physical geography, if not on the extent to which each topic should be studied, there is much less agreement on the content of human geography. The five major physical topics seem to have been accepted at an early date. In contrast, the content of human geography seems to be the result of a process of trial and error which is still going on. There is little argument about what topics should be studied, and which should not; for while the great geographers of the past and the more active ones of the present have very definite and quite different ideas, most writers have played safe by keeping to the shortest list of accepted topics.

If physical geography did not exist, it would be possible to sit down and create a scheme of operations in an afternoon by thinking out logically which topics of the physical world ought to be studied. Reading the works of some of the older human geographers, one gets the impression that they did this for human topics; and each came to a different conclusion. Working in the dark, this was probably inevitable, but it was essential groundwork which makes the work of the modern human geographer infinitely easier. Even if he does not agree with them, at least he has been saved the task of exploring and rejecting certain possibilities.

A survey of the content of human geography, then, is very much a survey of what a few men have done and are still doing. The topics mentioned below are those which human geography either has studied or is currently studying, but they are not necessarily the topics which human geography ought to study.

Fleure[5] arranged the topics of human geography into three groups, Life, New Life, and Good Life. Food is essential for life, and through nutrition the topic 'Life' could be extended to cover all aspects of economic geography. The topic 'New Life' involves the study of sex and societies, beyond what is currently accepted as social geography, while 'Good Life' studies the art and the general culture of the people concerned. Fleure also had the idea of classifying parts of the world as regions of increment, difficulty, privation or debilitation to point out how much surplus the economic activities provided for new life and the good life. This idea has not been developed, but if not applied too rigorously, it can provide useful themes in teaching regional geography.

In even more general terms a geographer, a sociologist and an ethnologist have divided the study of human geography into that of three themes. Le Play[6] in France, and then Geddes[7] in Britain, advocated the study of place, work and folk. More recently Forde[8] has called his study of simple societies *Habitat, Economy and Society,* but whatever the words, there is general agreement that one should study the place in which people live, the way in which they make a living, and the kind of life they have. Place is covered by physical geography, but in most of the human geography known to the present writer only certain aspects of the economy are considered, while the society, the way in which people live, whether they have a good or bad life, is hardly considered at all.

A warning is essential here. Le Play, and some of the textbooks which have borrowed heavily from Forde, state that the place decides the type of work, and the type of work decides the type of society; in short that people are conditioned by their environment and have to work and live as it dictates. This crude determinism will be considered later, but its existence needs to be pointed out here. The three topics of habitat, economy and society can be studied without necessarily implying that there is always a rigid connection between them.

De la Blache[9] went into more detail about the content of human geography. The writers mentioned above seem to be concerned with regional geography, but De la Blache was more clearly interested in the variations of certain phenomena from one part of the world to another. Thus he advocated the study of seven major topics:

World population
Groups of people
Tools and raw materials
Food production
Houses and settlement
The stage of development of the particular civilisation
Transport and communications.

Again we see the emphasis on how man makes a living, but while De la Blache was conscious of the many different societies and civilisations at different levels of technological achievement throughout the world, he showed less interest in the spiritual and artistic features, and concentrated on such things as tools and houses.

Brunhes grappled with the problem in such depth that it seems he felt two different approaches were necessary to give a complete cover in the geographic study of man.[10] Firstly, Brunhes arranges his material, in order, from basic needs to sophisticated organisation:

Prime necessities: food and drink, shelter and clothing.

Exploitation of the earth: farming, mining and manufacturing.

Economic and social geography: the study of groups and societies because people live together for reproduction, division of labour, distribution of land.

Political geography: the study of the relationships between these groups, societies, states and countries.

Secondly, in another arrangement, Brunhes pins down the actual objects of study, starting each of his three major groups with the phrase 'Facts connected with'—

The unproductive occupation of the soil: houses and settlement, highways (really all transport and communication).

The conquest of the plant and animal world: fields (i.e. arable farming), domesticated animals (pastoral farming).

The destructive occupation of land: mineral exploitation, destruction of plants and animals (e.g. forestry and trapping).

Not only does Brunhes pin down his facts very precisely, he also makes it clear that anything which is not directly connected with the physical environment is going to be left out of his work. While this may be considered restrictive, it is definite, and certainly avoids the danger of having the reader *assume* that there is some connection with the physical environment in a work which goes on to consider such things as marriage customs and political opinions.

Another French geographer, Demangeon, is briefer but nearly as strict.[11] For Demangeon the content of human geography includes modes of life in the climatic regions of the world and techniques of making a living, hunting, fishing, farming and trade. Equally important are population density, distribution, limits, migrations and types of urban and rural settlement. In some of these earlier arrangements, some topics may be missing because they were not important human activities at the time. Certainly the content of human geography must change as the activities of man on the face of the earth change. But other topics which certainly did exist and which have long been the subject-matter of geography are included in the more general phrases. Thus the last group of Brunhes and Demangeon's techniques of making a living include manufacturing. This is a problem inherent in classifying and defining. One seems to end up either with a long, complete, detailed list or a definition so neat and succinct that it needs a book to qualify it.

The arrangement put forward by Emrys Jones[12] recently strikes a realistic balance. While the order of topics is different again, most of them are familiar and generally accepted: population growth (distribution, density and structure); divisions of mankind (racial, ethnic, political, national, cultural, linguistic and religious groups); migrations (all types); obtaining food; rural settlement (materials, site, pattern, and distribution); mining and manufacturing; towns and cities (sites, pattern, groups of people in cities); communications.

A useful exercise is to tabulate these lists of topics which have formed the contents of some of the leading works on human geography, in order to reveal the nuances of thought and arrangement. However, in broad outline the only general agreement is that man's economic activities form the hard core of study. The basic processes whereby communities (rather than individuals) produce food and the other necessities of life are obviously of prime importance. Ignoring such finer divisions as distinguishing between the production of food or raw material on a farm, the major activities are fishing, forestry, farming, mining, manufacturing and transport, each in the widest possible meaning of the word each time.

From this point opinions obviously diverge. Even the attention given to the topic of population varies from the concentration on groups by Fleure, which we think of as sociology today, to the much more comprehensive studies envisaged by De la Blache and Emrys Jones, which extend beyond geography into demography and anthropology. Further on, study of such things as clothing, types of houses and tools seems to be quite optional, fascinating to some geographers, but beneath the attention of others. Finally, while political geography is an established discipline, it exists mainly by itself in books devoted completely to that topic, and receives less attention in the general and regional geographies. Thus consideration of the cultural, artistic and political nature of societies is very much on the frontier of human geography; or in backwaters long bypassed by the frontier, according to one's point of view.

One can consider many criteria for the selection of the content of human geography without coming to any cut and dried conclusion. The only assumption one can make with any safety is that selection from the sum total of phenomena on the earth's surface is absolutely essential to make the work manageable. There can be various starting points, such as the fact which gives rise to geography in the first place, simply, that one part of the world is obviously different from another. So we pose the questions, how is it different and what things are different? This will certainly give a start, but one is liable to reach that uneasy stage of passing from topics which are obviously important and worthy of study, like man's economic activities, to superficial and trivial topics such as the shapes of street lamps in different parts of the world. The problem then is where to stop.

From an examination of the works mentioned above, one might suspect that another approach had been used, starting out to study man's basic requirements of food, shelter and sex throughout the world. Farming and settlement certainly receive fair attention, but only in the works of Fleure and Brunhes do we get a tantalising hint of sex. While these are vital topics, correctly they are studied elsewhere, by sociologists, anthropologists and so on, and blindly to pursue them through geography could be most misleading for they do not necessarily show regional variations from one part of the world to another.

Other possibilities are suggested by contemplation of the map and the landscape. Geographers certainly do study things which are not visible on the ground, and maps show such things as parish boundaries which are not visible in real life; but the need to make maps and the urge to study geography stem from the same fact that one part of the world looks different from another. This results partly from the physical features and partly from what man has done and is doing in that physical environment. Relief and vegetation are the most important features in the appearance of the physical landscape, although exposed rocks, climate and soils are vital factors in causing this appearance. In the humanised landscape the crops, roads, walls, houses, factories all make their infinite variety of impressions, but two points stand clear: firstly, that there are so many more types of things to consider in the human field, and, secondly, that man's physical changes of the natural landscape are the most obvious, easiest to record, measure and map, and are the direct result of his attempts to make a living, or to provide himself with food, shelter and relaxation.

Thus another approach to the problem of what to select and include in human geography is by way of man's visible and concrete additions and changes to the earth's surface. Again we must bear in mind that there is little geographic value in studying some concrete phenomenon which is the same throughout the world, or in studying something which does vary from place to place but which is not of vital concern to the people in those places. For example, geographers pay scant attention to footpaths, which exist throughout the world, but often are concerned to know about roads in different countries. At first roads might seem to be as obvious, ubiquitous and uniform as footpaths, until closer study shows that the economy of a country is retarded by the desperate shortage of roads, that the gridiron pattern of the western USA means longer journeys than the radiating pattern of Europe, that eight-lane highways, and tracks with passing places are regional characteristics of parts of the USA and Scotland respectively, and that the lack of a tarred surface on the single road in and out of the country could mean disaster to a country like Zambia.

This is the stuff of geography. Such things are concrete, vary from place to place, and are of vital concern in the daily problem of making a living. In contrast field boundaries, styles of windows, dress, and designs painted on barns may vary enormously and make a vivid impression on the traveller. But they are likely to be matters of tradition or fashion which please the local people while having no significant effect on their work, and which they might be ready to discard on a whim. They might form sound subject-matter for books on architecture, national costume or native design, but the geographer usually finds it difficult to make them part of the fabric and structure of his work.

Obviously the scale of the study and the size of the area are important. In a study of northern England one glassblower in St Helens doesn't count: but in a study of Torrington or an Irish puckoon, the three men in the new glass works which is the only industrial activity obviously do. Similarly, even in big cities, geographers study shopkeepers but not actors—until the rare occasion when Hollywood or Cine Citta is too obvious to ignore. The signs are that these points have not been thought out clearly beforehand (which is not really necessary), but also that individual geographers have rarely established general procedures when they have successfully tackled particular problems. The tendency is to play safe and do only what others have done. Yet here we are on another frontier, where the content of human geography is not confined for all time, and there is room for experiment, change and extension.

A nice distinction must be made between the fact that much of the content matter of geography is material common to many disciplines, and the allegation that geography just repeats work done better in other places. Close examination reveals that surprisingly few of the sixteen topics most commonly touched on in geography exist as subjects or disciplines in their own right. Study of the physical topics and of population are recognised branches of learning outside geography but most of the work written about human topics is done only by geographers.

The idea that geography simply repeats work done in other subjects has some validity in the case of physical geography, superficially at least. Each of the five topics is a recognised discipline, with its own aim and method:

Topic	Discipline	The study aims to discover
Rock	Geology	Formation of the rocks, including past erosion history. Nature and arrangement of the rocks.
Relief	Geomorphology	Types of landform. Processes of erosion and deposition. Principles.
Climate	Meteorology	Elements of, and factors in, weather and climate. General principles and laws.
Soil	Pedology	Types of soil, factors in their development.
Vegetation	Botany	Types of vegetation, structure and growth.

This table is very simplified, but not to the point of distortion. For example, geology may be further subdivided into petrology, mineralogy and palaeontology, and linked with other subjects such as geophysics or crystallography. It would be presumptuous to try to define these disciplines each in one box in a tiny table when a book is needed to try to define geography, but the essential points to be underlined here are that these five disciplines are natural sciences and as such are interested in studying types, discovering general laws and investigating the nature of the topics of study. They study processes and systems, often in historical depth, but in establishing laws and general principles are rarely concerned to describe and explain the complete, world-wide distribution of those phenomena.

At this point where their interest stops, the work of the geographer begins. He may call himself a physical geographer, a climatologist or a bio-geographer, but what makes him a geographer is not his label, but the type of work he does in studying the actual distribution of these phenomena on the earth's surface. Geography was defined as 'the science of distributions'[13] by one who took a very specialised view of the subject. Here one must admit that the allegations that geography is no different from such disciplines as geomorphology are well founded, for very few self-styled 'geographers' in fact do what in theory they should—study world distributions and spatial relationships. General physical geographers too often turn out to be geologists, geomorphologists or pedologists-cum-botanists wearing the wrong label. There are maps of the distribution of rocks, relief and vegetation throughout the world, but not complete maps of landform types and detailed maps of soils. If five sets of such maps were complete and available then we would be a long way towards having genuine systematic general physical geographies.

Some geologists, pedologists and botanists are concerned with the areal distribution of their objects of study, but we have yet to see a work of general geomorphology which deals with the world-wide distribution of types of landforms. In the table (p. 23) the word meteorology was used for the discipline connected with the geographical topic of climate. Here the author would stress that climatologists such as Miller[14] and Kendrew[15] were *de facto* geographers in that after dealing with types and principles, they went on to describe and explain the world distribution of climates. In contrast, some physical geographers are *de facto* geomorphologists who confine themselves to types and principles, and do not go on to consider world distribution. The world distribution of rocks, types of landforms, soils and vegetation may exist fragmented in the regional geographies of a great number of men who have concentrated on the actual geography of the different parts of the world.

Here one is in the position, again, of seeming to say what 'geography' ought to do, instead of simply describing what geographers do. But in physical geography the geographical gaps are so obvious, and the criticism that geography does not exist as a separate discipline is so well founded, that it seems valid to point out *a priori* the theoretical difference between the natural sciences and physical geography. There are very few real general physical geographers, but many geologists and geomorphologists, in particular, have been labelled geographers.

In the topics which are usually the concern of economic geography, a distinction must be made between the work the people do and the things they produce or the objects with which they work. Otherwise one can become very confused when trying to distinguish the other disciplines from which geography 'borrows'. For example, in the topic of commercial sea fishing which is normally included in economic geography, the objects or produce, the fish, are studied by the zoologist. But is there a separate discipline which studies fishing, the work of which is duplicated by the economic geographer?

Topic			*Related*	*Study of the*
Activity	*Object*	*&*	*discipline*	*activity*
Trapping	Animals ⎫			1 ⎫
Fishing	Fish ⎬		Zoology	
Pastoral farming	Animals ⎭			4 ⎪ see the
Forestry	Trees ⎫		Botany	2 ⎬ notes in
Arable farming	Crops ⎭			4 ⎪ the text
Mining	Valuable minerals		Mineralogy	3 ⎪ below
Manufacturing	Goods		? ?	5 ⎭

(1) There are books such as Morgan's *The World's Sea Fisheries*[16] but it is not clear whether these constitute a recognised discipline, or whether the authors think of themselves as geographers. (2, 3) Forestry and mining are studied in separate colleges and in departments of universities, but much is research as natural science,

practical application as technology, and vocational training. (4) Farming or agriculture is certainly studied widely in many places. But the characteristics of forestry, mining and farming which seem to set them apart from the natural sciences discussed earlier are the much greater emphasis on vocational training, and the apparent shortage of general academic works on which the layman and geographer could draw. (5) In the case of manufacturing, much study has been done on individual firms, industries and processes, and training in business management is well established in some countries, and perhaps the point here is that a mass of material waits to be organised into a discipline.

In the light of this, one could make a modest claim that far from copying and poaching, geographers have been the pioneers in the systematic study of most of these vital topics. Briefly, economic geography studies the nature of these activities, the distribution of them throughout the world and the reasons for the special patterns of distribution, and the amounts of commodities produced. There are some sub-divisions into finer detail, and here the pioneering has been done. In addition to Morgan's work there are general geographies concerned mainly with one topic such as farming, mining, manufacturing or communications.[17]

Some economists are interested in these topics as well as the economic geographer. Thus a topic such as farming may form the content of three disciplines, and be studied from three points of view. The agronomist is interested in method, the economist in production, costs and trade, the geographer in the locations of the different types of farming, both in place and in relation to physical and economic factors. Acquaintance with economics, however, strengthens the impression that economists only touch on these activities among many others and are not concerned to study them in a systematic way. This strengthens the argument that geographers do much of the original work in this field. Only geography considers the activities of fishing, farming, mining and manufacturing in detail, in the actual places and under the real conditions in which they take place.

Geographers, then, put the emphasis on the geographic distribution, but in doing so seem to lose sight of many of the economic conditions. In fact one gets the impression that geographers do not borrow enough from economics because only the most straightforward ideas of supply and demand are used in regional texts, and very few of the more advanced and realistic concepts in general geographies. Such geographers as McCarty[18] stand out for their application of involved economic concepts in the explanation of the location of man's economic activities. In contrast, quite detailed general economic geographies largely ignore price fixing, cartels, freight increases and all the seemingly 'double-dealing' in finance which greatly influence the type of farming, the location of plant and the continued existence of theoretically 'uneconomic' enterprises.

While geographers have a different aim from geomorphologists and economists, and use the material in a different way, they need to have a fair knowledge of these other disciplines. One aspect of of the nature of geography in the mid-twentieth century is the serious lack of balance in the position of geography in relation to geomorphology and economics. The majority of geographers, students and school pupils, from the evidence of their books, special studies and syllabuses, appear to be very well grounded in geomorphology and the other physical sciences. This makes all the more disturbing the fact that they appear to be so ingenuous when it comes to economics, and are satisfied with ideas and statements which at the same level in geomorphology they themselves would criticise as superficial and worthless. In most of the geography courses in schools, colleges and universities known to the author, there are separate sections going into detail in geomorphology, geology, climatology and the like, even into mineralogy; but he has yet to find one which maintains a balance and goes into equal detail in economics. A knowledge of economics is essential for advanced geography, and is of more value to the pupil who goes straight to work from school than is the knowledge of glacial overflow channels and the intertropical front.[19]

Topic	Discipline	Branch of geography
Transport and communications	Civil engineering?	Economic
Rural settlement	?	Social
Towns and cities	Architecture	Urban
Population	Demography	Social
Political units	Political science International relations International law	Political

Turning finally to the rest of the topics, again we find two familiar points. First, it is difficult to name a separate discipline which deals with some of these topics as its main function. Hence the question marks and the inclusion of architecture. Some architects are very interested in towns and cities as whole units, and in town planning, but these are not the main concern of architecture. Secondly, where disciplines do exist, again they study types, and the real positions and distributions on the ground are seldom important. Types of rural settlement, types of cities, types of government, types of societies; only demography comes close to geography in studying often the numbers, growth, composition and movements of people by areas.

The study of people, in fact, is broken into many more disciplines than the table suggests, including demography, sociology, anthropology, ethnology and so on. Moreover, geography itself here shows a greater fragmentation than anywhere else. The geographical study of these topics is usually called either human or social geography, but there is also urban, population and political geography. Human geography usually includes economic geography, while social geography is usually restricted to settlement, towns and population. The main reason for this fragmentation seems to be that the social and urban geographers, like the economic geographers, are genuinely breaking new ground. As no disciplines exist which study these topics systematically, the geographers are having to specialise very narrowly in order to do three jobs: to write the history, establish the principles, and study the geography of the phenomenon in question.

Certain vital points have emerged from this brief review. Many of the subjects from which geography is said to borrow just do not exist. As a result, human geography is breaking much new ground. If one is prepared to encourage human geographers to become town planners and demographers, it seems illogical to criticise physical geographers for reverting to geomorphology. However, there is much repetition in these physical branches, while there are serious gaps in our knowledge of human behaviour. In particular, geographers proper need more training in economics and sociology, and less in the physical sciences. Some branches of geography, which exist in theory, are covered at the time of writing by only one pioneering book. As titles of books can be so misleading, and the academic posts in geography often differ from the kind of work they do, one must examine the contents of the book and the actual work of the man to decide whether this falls within geography proper or within one of the very closely connected disciplines.

At present the content of human geography consists of the topics of all methods of producing food and raw materials, manufacturing, rural and urban settlement, communications, population, and political boundaries and units. Stated so baldly this appears to be a very odd mixture, haphazard in choice and arrangement, but the earlier pages have suggested how this came about. While this content may well change in time,[20] however, certain underlying principles, which have also been examined, guide the choice; guide it but do not restrict it. Human geography selects phenomena for study according to five guiding principles. The subject-matter should either be visible on the ground or capable of being shown on a map. Given this, it should be worldwide in occurrence, but may vary from place to place. The subject which varies from place to place should have connections with other phenomena which vary. The phenomena should be significant to man; and they should suggest logical arrangements in space.

Geography exists as a study because phenomena vary from place to place. It has dignity as an academic discipline because the connections between varying phenomena can be studied and explained. It limits its field of enquiry to the five topics of the physical environment and the three major topics of human life, economic activity, settlement, and social organisation. It has value in that it selects from everything on the earth's surface those aspects which, in order for him to survive, are most significant to man.

1. Lukermann, F., 'Geography as a formal intellectual discipline, and the way it contributes to human knowledge', *The Canadian Geographer,* vol. VIII, no. 4, 1964, p. 167
2. Lukermann, op. cit., p. 168
3. James, P. E., in Cohen, S. B., *Problems and Trends in American Geography,* Basic books, 1967, p. 5
4. The lithosphere, atmosphere, hydrosphere and biosphere.
5. Fleure, H. J., 'Human regions', *Scottish Geographical Mazagine,* vol. XXXV, 1919
6. See Freeman, T. W., *A Hundred Years of Geography,* Duckworth, 1961, p. 87
7. ibid.

8. Forde, C. D., *Habitat, Economy and Society*, Methuen, 1956
9. de la Blache, P. V., *Principles of Human Geography*, Constable, 1926
10. Brunhes, J., *Human Geography*, Harrap, 1956
11. Demangeon, A., in *Geography in the Twentieth Century*, Ed. Taylor, G., Methuen, 1957, p. 82
12. Jones, E., *Human Geography*, Chatto & Windus, 1964
13. Marthe, quoted by Taylor, op. cit., p. 66
14. Miller, A. A., *Climatology*, Methuen, 1953
15. Kendrew, W. G., *The Climates of the Continents*, OUP, 1953
16. Morgan, R., *World Sea Fisheries*, Methuen, 1956
17. e.g. Dumont, R., *Types of Rural Economy*, Methuen, 1957; Miller, E. W., *A Geography of Manufacturing*, Prentice Hall, 1960; Sealy, K. R., *The Geography of Air Transport*, HUL, 1957
18. McCarty, H. H., *The Geographic Basis of American Economic Life*, Harper, 1940; McCarty, H. H., and Lindberg, J. B., *A Preface to Economic Geography*, Prentice Hall, 1960
19. See, however, Chisholm, M., *Geography and Economics*, Bell, 1965
20. For the content of human geography see also:
    Febvre, L., *A Geographical Introduction to History*, Routledge, 1925
    Bryan, P. W., *Man's Adaptation of Nature*, ULP, 1933
    Sorre, M., *Les Fondements de la Géographie Humaine*, Armand Colin, 1952
    Perpillou, A. V., *Human Geography*, Longmans, 1966
    Unstead, J. F., *A World Survey*, ULP, 1955
The tendency now is for specialisation in one branch of human geography, with more subdivision in that branch, e.g.
    Clarke, J. I., *Population Geography*, Pergamon, 1967
    Beaujeu-Garnier, J., *Geography of Population*, 1966, and *Urban Geography*, Longmans, 1967
    Clarke, C., *Population Growth and Land Use*, Macmillan, 1967

3

IS CERTAIN SUBJECT-MATTER
UNIQUE TO GEOGRAPHY?

Although for the moment we may be begging the question of what geography is and what geographers are, the emphasis in the last chapter was on the subject-matter of studies written by those generally accepted as geographers. An attempt was made to avoid working out the content of geography from first principles. From the considerations in Chapter 2 it seems fair at this stage to make certain comments on content or subject-matter. The phenomena studied by the geographer are those distributed over the earth's surface, and in one way or another are necessarily connected with that real surface. At the moment, geographers as a body include in their study inanimate phenomena as well as animate phenomena, flora, fauna, and man.

The subject-matter is narrowed down to what is important to man, while the study of man is restricted to those of his activities and creations which affect his physical survival on this planet.

Since no consideration has yet, at this early point, been given to either the methods or purposes of geographic study, it has not been established that geography has a claim to be a separate discipline on the grounds of either its subject-matter or objects of study alone. Clearly the basic subject-matter of human geography is shared with disciplines such as history. The geography of manufacturing and the economic history of manufacturing deal with the same facts in a different way. Very often a student of geography finds it very difficult to distinguish between, say, the geography of farming, the economic history of farming, and an historical geography of farming. It is hoped the difference will be made clear in the section on method, but the point here is that of common subject-matter.

It is easier to distinguish between the methods and aims of the physical geographer and the natural scientist, but again there can be no doubt that they study the same phenomena. The geologist, mineralogist, palaeontologist, geomorphologist and geographer are all interested in rocks for different reasons. Geomorphologist and geographer are named separately quite deliberately.

While certain phenomena are obviously the subject-matter of more than one discipline, this does not mean that they are of no significance in helping to define the disciplines. The young student probably needs to get his ideas of subject-matter in order, as part of his general theoretical framework. More important, even when the subject-matter is the common property of several disciplines, the author finds it hard to understand those who assert that methods are the only important considerations because in the end our methods are to give us understanding of certain *things*. If spatial analysis is the only important consideration, we may as well play with matchsticks on the table, instead of studying landforms, crops, settlements and population. Marthe may have made a succinct definition of geography when he called it 'the science of distributions', but the question still remains—distributions of what?

Possibly, if it seems that no subject-matter is unique to geography, and if we accept Marthe's definition, geographers can study *any* phenomenon. Some guides for limiting such infinite possibility were suggested in the last chapter.

Even with these guides, however, the choice of subject-matter still seems to be pragmatic. This can be simply dissatisfying to some, and a matter of ridicule for others. In an introductory work such as this, and at a time when what we call geography is growing and changing so quickly, we may have to leave it at this.

On the question of content, three points are clear. First, that individual topics or phenomena such as landforms and population are not unique to geography; but at the same time are a vital and necessary part of it. Second, geographers have at least developed and furthered the study of certain topics. Thus what was once unique to geography has now grown until it is common property of several disciplines or a separate discipline in its own right. Accepting again the danger of misapplied labels, what started as physical geography and was developed so rapidly by W. M. Davis has become the separate science of geomorphology. Similarly, in their need for original data, social geographers in particular have developed the study of the subject-matter of rural settlement, towns and cities, communications and population. As disciplines such as topography, rural studies, environmental studies, regional studies, demography, and the many overlapping branches of social geography proliferate, it is easy to confuse cause and effect and to think of geography as not directing its attention to any subject-matter in particular.

Thirdly, the list of the contents of geography must be open ended. Not only may man's activities on the earth's surface provide phenomena relevant to the geographer's work in the future, but also geography may not yet have developed to that stage where it has taken account of all the relevant phenomena which exist now. The history of geography reveals this constant inclusion and exclusion of phenomena, or at least the changing importance attached to the study of phenomena such as soils, vegetation and settlement.

As to the first point, that individual topics are not unique to geography, there are many authorities who believe that geography does have unique subject-matter or objects of study when several topics are taken together. Thus Lukermann[1] accepts regions as geography's objects of study almost without question. This is not the place to go deeply into the nature of regions,[2] but one or two attitudes to regional geography are relevant to the claim to unique subject-matter. Those who support the theories of regions state in different ways that it is the combination of several phenomena on the earth's surface that makes one part different from another. If this is so, then geography exists as a separate discipline with its own unique objects of study when attention is directed to these regions. The region may be taken simply as an area of roughly homogeneous relief, climate, land use and so on, or as a highly complex functioning unit of heterogeneous parts. Hartshorne[3] takes exception to the extreme view of the region as an organism, but without going to this extreme it does seem possible to define regional geography by its subject-matter of different types of regions where two or more phenomena are studied in combination. In spite of the misleading name of the new discipline, regional science, regional geography in this sense has unique content as well as its own methods and aims.

Pursuing further this idea of the combination of phenomena on

B

the earth's surface, the visible landscape which is the sum total of the interaction of those phenomena has been claimed as the object of study of geography. This idea has waxed and waned several times, in Germany and France with reference to landscape, in the USA to the cultural landscape, and recently in Britain to the landscape as the object of study of the discipline of topography. A fine distinction must be made between studying the landscape, the combination of many phenomena, and studying the interaction of phenomena on the earth's surface. The claim of certain geographers to study the interaction of phenomena will be considered later.

The trend in regional geography at present is to concentrate on functional regions and single-topic regions.[4] This fact, combined with remarks of Borchert and Chisholm, suggests another answer to our question. Borchert writes: 'The committee considered the overriding problem of geography to be the understanding of the vast system on the earth comprising man and the natural environment.'[5] Chisholm,[6] in criticising the new jargon of general systems theory, insists that geographers have been studying systems for a long time under other names. He gives examples of how our concepts of the cycle of erosion and of functional regions can be re-worded in systems terminology.

A slight digression into systems theory may be helpful here, for the key word in the study of our problem is now 'system'. While Chisholm claims to be sceptical about general systems theory, Borchert, Berry[7] and others believe that it will be most helpful in formulating our concepts, developing new ideas, and setting them out in such a way that ideas can be transferred easily from one specialism to another.

A system is any group of phenomena which have a functional relationship. One valid criticism is that the terminology of the theory must be so general that people still find it easier to grasp the concept in concrete, particular terms. Thus as geography students we are familiar with agents of erosion using debris to change landforms in time; with the dynamic wind system producing periodic changes in temperature and rainfall; with the closely related system of climate, vegetation and soils; with raw materials, factories, labour and markets, forming a functioning whole dependent on communications; with a town serving and being served by surrounding land.

General systems theory seeks to generalise the functional

relationships and processes which have been recognised in these particular systems, and the concepts behind them. It is then hoped to establish models which can be applied to real systems in science, engineering, geography and so on. Some of this may not be a new idea, but the idea of having some kind of central clearing house, into which all the features of special systems from different disciplines can be fed, and then from which the valuable general concepts can be distributed to all disciplines, seems to be valuable. As more and more people specialise ever more deeply, there is a need for a service which collects and makes available ideas common to seemingly widely different studies in order to avoid duplication of work, and to make available concepts which will save thought and so speed up some work.

'The existence of laws of similar structure in different fields enables the use of systems which are simpler or better known as models for more complicated and less manageable ones.'[8] Moreover, laws from one field can suggest the existence of similar laws in another field. The best known example in geography is the adaptation of laws of gravity to urban fields and to traffic flow in the form of the gravity model.[9] However, Chisholm goes on to insist that general systems theory must develop a critique of analogy and a theory of simulation in order to avoid superficial or even dangerous analogy. He sees the greatest value lying in the chance of general systems theory integrating both the separate branches of geography, and geography itself with other disciplines. Thus a general systems theory worked out, among others, from:

A. Physical
B. Engineering
C. Computing } Systems
D. Social
E. Biological

could be applicable to, say, cybernetics B. C. D. and geography A. D. E.

Borchert emphasises other values of the theory. By adopting the more generalised terminology, geographers will be able to put their data into computers more easily, although Borchert is not explicit on this point. Greater familiarity with systems certainly will help students of geography to visualise such things as functional regions more readily, and to comprehend limited topic functional relationships without necessarily locating them in a particular landscape. An incidental advantage, which seems to have great potential value, is that familiarity with general systems theory should also tend to correct geographers' bias to average conditions or the 'steady state' conditions. Borchert reminds us that we tend to concentrate too much on average, seemingly static conditions in maps and graphs of climate, traffic flow, population, etc. Attention to systems will necessarily demand more concentration on the dynamics, the periodic variability and long-term change, on functional relationships and interaction.

Returning to the examples of familiar systems at the beginning of the digression, we see that these are still separate systems, and geography can not claim exclusive interest in any one of these. Geomorphology, climatology, ecology, economics and other disciplines each study one or more of the systems for their own purposes. While these systems are not in the geographer's unique province, they do better satisfy the need for proper subject-matter, and form more easily identified objects of study. However, Chisholm demonstrates that all these systems are open systems, and the separation of any two must in the end be an arbitrary matter for the purposes of defining an area of study. While investigating causes and effects in his examination of determinism, Martin[10] comes to the same conclusion in different words for different reasons. But both Martin and Chisholm lead us to the same conclusion voiced by Borchert, namely that there is one vast system, comprising the whole earth's surface, which innumerable geographers in their many ways attempt to study.

Chisholm's conclusion that general systems theory is just new jargon for old ideas seems appropriate here. When Borchert writes 'the understanding of the vast system on the earth comprising man and the natural environment' in 1967, one is reminded of Hartshorne's phrase 'world comprehension', written in 1939. However ambitious the project, different writers at different times have come to the same conclusions. First, that anything less than the whole world is not the unique subject-matter of geography. Second, that the object of study of geography is the narrow sphere just above, below and on the earth's surface. Third, that geography is the only discipline which attempts systematic study of that real surface; and fourth, that while in the eyes of some the fact of only one object of study will prevent geography being a science, that object is unique in its supreme importance to man.

1. Lukermann F., 'Geography as a formal intellectual discipline, and the way it contributes to human knowledge', *The Canadian Geographer,* vol. VIII, no. 4, 1964, p. 167
2. Minshull, R. M., *Regional Geography: Theory and Practice,* HUL, 1967
3. Hartshorne, R., 'The Nature of Geography', *AAAG,* Lancaster, Pennsylvania, 1939, p. 256
4. Minshull, op cit., p. 38
5. Borchert, J. R., in Cohen, S. B., *Problems and Trends in American Geography,* Basic Books, 1967, p. 268
6. Chisholm, M., 'General Systems Theory and Geography', *Transactions of the Institute of British Geographers,* no. 42, December 1967
7. Berry, B. J. L., 'Approaches to Regional Analysis', *AAAG,* vol. 54, March 1964, p. 2
8. Quoted by Chisholm, op. cit.
9. Cole, J. P., and King, C. A. M., *Quantitative Geography,* John Wiley, 1968, p. 503
10. Martin. A. F., 'The Necessity for Determinism', *TIBG,* no. 17, 1951, p. 1

4

THE GEOGRAPHICAL APPROACH

Lukermann begins his article in *The Canadian Geographer* by stating that neither content nor method is unique to geography, and that geography, like any other discipline, is defined by the questions it asks.[1] As this statement is at variance with the commonly held view that method, the way of dealing with subject-matter, is the distinguishing feature of geography, it is necessary to try to sort out some terms.

Some writers use the word method to include both the general approach to the subject-matter and the particular devices of using such things as maps and computers in studying phenomena. In this chapter, the word techniques will be used for these particular practical methods which again are not necessarily unique to geography. Local historians use maps, and many kinds of specialists use statistical techniques, but some mention of their place in geography will be made later. The main concern here is to consider whether there is a proper geographical method or approach to that subject-matter which other disciplines study by different methods for their own purposes.

The results of this consideration may help one to agree or disagree with Lukermann that method is not unique, but there is also his implication that the questions raised in, and the purpose of, geography are different from method. Here again, different writers make different assumptions; some, simply, that studying the combination of phenomena in place on the earth's surface is at the same time both the method and purpose of geography. Inevitably, then, some mention of purpose will have to be made here, but the main examination of this point will be made later.

In examining the question of a geographical approach, it seems more profitable to proceed along theoretical, deductive lines. In addition to the confusion caused by the neglect of many writers to separate techniques, methods and purposes, the empirical, inductive method would more easily put one in the position of trying to discern a geographic method in the work of those who did not adopt a special approach. The author makes this statement from the conviction that many students believe that they are writing geography provided that they are writing about rocks, climate, farming, towns and so on; emulating people who have written what purport to be geography books, but in which, on theoretical grounds, one can find no special approach beyond the cataloguing of phenomena in certain parts of the world.

An attempt will be made to deduce the method and approaches of geography from the hypothesis of the position of geography in relation to the other disciplines which study parts of the external world. One common view of the position of geography is that it cuts across the subject-matter of all other disciplines at right angles. In this view geography covers a bit of geology, a bit of biology, of physics, economics, architecture and so on, and is a hodge-podge of odds and ends, very useful for school projects, but not capable of existing as a separate discipline.

Another common view is that geography is an odd but useful subject which forms a kind of bridge between history and natural science. Well-educated people who are aware of the different approaches of the historian and natural scientist realise that the geographer coincides with neither, and too often conclude, therefore, that his approach must combine those of the other two. In this fact we see the beginnings of both the idea that the geographer is a jack-of-all-trades and of the damaging division of geography into physical and human halves. Some people further waste their time debating whether geography is a science or an art. At least since the time of Kant it has been clear that geography is neither of these. There are not two ways of studying reality but three, and the third member of the trio is not a thin bridge between the other two, but is of equal stature and individuality. Emmanual Kant (1724–1804), Alfred Hettner (1859–1942) and Richard Hartshorne have in turn clarified and emphasised the position of geography as one of the three fundamental ways in which man studies reality. According to them, there are three ways of studying reality: The relationship of similar things (natural science), Development in time (history), Arrangement in space (geography). Kant, as a philosopher rather than a practising geographer (although he lectured on geography for forty years), firmly believed that geography is the most important of these three, because it results from man's first and direct contact with the complex reality outside his body. However, there is nothing to be gained in advocating a hierarchy of these three divisions. In fact, as a discipline, history has existed longest, systematic work in natural science dates largely from the Renaissance and has boomed in the last 150 years, while geography is now coming into its own.

In the list above, the words history and geography are not strictly correct, and Hartshorne has elaborated on this.[2] Using the word science in its correct sense of 'knowledge', Hartshorne distinguishes the systematic sciences, the chronological sciences and the chorological sciences: studies of things in isolation, in time and in space. Systematic sciences include: natural science, e.g. biology, chemistry, physics; social science, e.g. sociology, demography, anthropology. The chronological sciences include: palaeontology, prehistory, history. The chorological sciences include: geophysics, geography, astronomy.

In the first group the object or phenomenon forms the unit of study and is isolated for that study of its nature and behaviour in response to natural, physical laws. Physicists study mechanics and hydraulics in isolation, not in place, mixed up with everything else; as the geographer studies the application of shaduf and water-wheel or turbine. The demographer may have an abstract idea of the numbers and ages of people in a town, the rates of population growth and migration, but the geographer must have the more complicated picture of the factories, shops, schools, houses, parks and roads as part of the environment in which these population statistics become people with stiff legs and bad colds trying to make a living or save up for colour television.

The second group may surprise a few palaeontologists and historians to find that they are in the same trade. The point is that a complete study of the development of our environment in time must turn to the evidence as it exists in different forms. What little evidence there is of the early history of the earth exists in the arrangement of the rocks, and in the fossils they contain. Between the time of the last deposition of sedimentary rocks containing fossils, and the time of the earliest written records, our only remaining evidence lies in the tumuli, barrows, strip lynchets, earthworks, standing stones, arrow heads and beakers of early man. In each case the unit of study is a period of time.

More obviously in group three, geophysics is concerned with the study of the earth as a solid body in space, and of its interior; geography with the surface of the earth, and astronomy with the universe beyond. The unit of study is an area of space.

Before setting out the three divisions clearly, Hartshorne points out that unless natural science discovers physical laws it is just description. Similarly, unless history demonstrates causal connections it is just a timetable of events, and unless geography demonstrates causal connections in space it is just an encyclopaedia. Following on from this, it seems that just as the discoveries of the natural scientist can be put to practical everyday use, and thus subconsciously we examine the usefulness of things, so the discoveries of historians and geographers can have practical value. The historian at times examines the rightness or success of past events and people, providing guidance for the educated person and thus, we hope, for all our leaders. Therefore the geographer might examine (as part of his study) the rightness or suitability of the land use in an area, the usefulness or potential value to man of another area. Such impartial studies, carried out by the side of more objective academic work, could become of increasing value in the world as the fast expanding population makes increasing demands on space.

The natural scientist must examine things as they are; there is no right or wrong, mistake or success in nature. The historian tries to reconstruct things as they were, but also to evaluate them. The geographer, above all, must see things as they are, and this is most difficult because the condition of man in some parts of the world is not as good and as happy as it might be, and not the way many geographers personally might like to see it. So the geographer could first see how things are; and then see actual or potential mistakes in land use and suggest possible developments. Sometimes this is not done because the geographer tries to be a natural scientist in two senses. First, some believe that man behaves as consistently as natural phenomena, and the only conclusion then is that whatever man is doing in a certain part of the world is the right and best possible thing to do there. Secondly, some try to be objective, and not to comment on what they describe and explain.

Rare is the geographer who says that people have made a hash of things in a given region, and then makes sensible suggestions for improvement. Fortunately there are a few of these rare men.[3]

In one short paragraph Hartshorne sums up how important it is for a geographer to be aware of the position of geography as one of three, and the disaster which results from thinking of it as one of two: 'students who cannot accept the particular characteristics empirically demonstrated as essential to geography, because they cannot understand that necessity, repeatedly attempt to change the subject to fit their view of what a science should be. The long history of such attempts demonstrates that their only effects are the personal frustration and professional unhappiness of those who try to fit a square peg into a round hole.'[4]

While stressing the separateness and individuality of the chorological sciences, it cannot be denied for a second that all disciplines are closely interconnected. In the first place they are responses to the common urge of men to study everything they perceive. In detail, of course, it is not a case of geography borrowing subject-matter from other disciplines, but of all disciplines having much subject-matter in common. Similarly many of the techniques of research, analysis, synthesis and presentation are common to all disciplines. Brunhes[5] gives a vivid example of similar content and method being used from two different points of view. The botanist, as a natural scientist, separates the cactus and aloe into different species, while he groups wheat, maize and rice together as cereals. The geographer, in contrast, as a chorographer, but still trying to arrange his knowledge in a systematic way, groups the cactus and aloe together because he finds them together in the desert; and he separates wheat, maize and rice as typical of different climates, economies and cultures in different parts of the world.

Perhaps part of the great strength of geography, and certainly its educational value in school, stems from the fact that geographers and geography teachers are well aware of the interconnection with other disciplines. This may be called 'borrowing' or 'poaching' by the uninformed, but the result is that all those who have studied geography go out of their way to think outside their problem or subject, and are not ashamed to learn from other disciplines as necessary.

Chapman[6] has developed this idea of the position of geography a little further in considering what he calls the dominant perspective of geography. This perspective or approach to the subject-matter focuses attention on 'the intersect of place, space, and time'. Chapman's article is valuable for this emphasis on place *and* space, and for bringing in the element of time. There is an increasing tendency in geography to isolate phenomena from the earth's surface, and to analyse spatial relationships, and the word space reminds us of this. Moreover, geographers study phenomena in place or in space at a particular stage in time, whether this be at the present time or at some definite time in the past as a re-creation of the historical geographer. However, Chapman's article by itself is not a sufficiently detailed exposition of the geographic approach or dominant perspective, because at the crucial stage in the argument space and time may well seem to have equal importance to the novice.

Chapman does go on to qualify this idea with other similar phrases, but not necessarily to clarify it any further. Thus he mentions the 'behavioural, historical-genetic and regional perspectives'. The first part on the dominant perspective is of help here, but a later section of the article, while separating method from techniques, lumps scientific method, general systems theory, systems analysis, model building and mathematical logic all as part of the method of geography. One would question, at least, whether scientific method is *part* of geographic method. However, Chapman's main purpose is to stress the value and importance of geography, and his other contribution to our present problem will be brought in later.

From Hartshorne, and from Chapman, we have this basic idea that geography is the study of the arrangement of phenomena in space, and particularly on the earth's surface, or in a definite place. Several deductions from this are possible, and the approach of 'geographers' has varied, and still varies, according to the emphasis on one or more of the deductions.

The first approach is simply to study the earth's surface, and to take account of everything found on that surface. There is some logic in this in that obviously the combination of phenomena in the landscape is not entirely haphazard, or random. Hartshorne[7] shows that many geographers have stated their method to be the study of the phenomena as they are actually combined on the earth's surface, as distinct from the scientists' method of studying each phenomenon in isolation (as far as possible). Geography with this basic approach has varied from detailed, but straightforward, description of the earth's surface, to works which attempt to give a full explanation of everything which has been observed and described. Many of the regional geographies written in the past forty years do little more than describe,[8] while men like Humboldt and Ritter (also Ratzel somewhat later) were the last to contemplate a full explanation.[9]

The danger of this approach, of course, is in confusing the study of the way in which phenomena are combined on the earth's surface, with the study of the earth's surface itself. So the earth's surface, or the landscape, or total regions, have become objects of study, and at times it is difficult to see how some 'geographers' have adopted any unique method in their description of landscapes. The landscape school of geography has been strongest in Germany and France, although Turnock[10] suggests that geography in most Western countries has passed through a similar phase. Quite clearly, at times regions have become objects of study rather than a method of selecting the arrangement of phenomena for study.[11]

A logical extension of this approach is to study the combination of phenomena within some limited part of the earth's surface. This may be called regional geography, and in particular this approach is seen most clearly in the study of formal or homogeneous regions. This may also be called the vertical approach by which a geographer studies the combination of several phenomena covering the same area or site. This approach involves studying the rocks, soil, drainage, relief, vegetation, climate, land use, communications, settlement, population and so on as they are superimposed, one on top of another.

There have been many variations on this theme to try to make the method as precise as possible. At the time when Herbertson[12] first published his work on natural regions there was some idea not only that regions of each phenomenon were sufficiently uniform, but that one type of relief coincided with one type of climate, one type of soil, of vegetation, of land use, culture and population density so that such regions could be defined easily. Having assumed this co-extension of phenomena, then causal connections were often described, but rarely proved. Later, particularly in the work of Whittlesey,[13] it was realised not only that phenomena are not so neatly co-extensive, but also that this is not a necessity for geographic study. With improving techniques and more detailed data, tests have been applied to measure the degree of correlation of two or more distributions, and the correlations can be ranked in order of importance.

At the same time there has been increasing awareness of the fact that not all phenomena which are found distributed over the same area are causally connected with other phenomena. In other words, there are coincidental correlations. Combined with this has been the growing insistence that on the earth's surface it is not necessarily a simple case of physical cause and human response, a fact which too many geography teachers do not appreciate yet.

This study of phenomena which are co-extensive and coterminous is often removed from the real landscape by people who still call themselves geographers. Thus the study of the connections between such things as population, religion, wealth, health, occupation, or between population, economic activities and urban fields is being divorced from the real landscape and seen as a matter of arrangement in space (neat, geometrical space) rather than in place (untidy landscapes where mountains, lakes and infertile areas distort the nice patterns).

In contrast to the second and third approaches, the method of the general geographer logically should be to consider the arrangement of one or more phenomena through the area relevant to his discipline, that is throughout the earth's surface. The phrase 'should be' is used for two reasons. First, because this is a logical deduction from the hypothesis of the position of geography, and second, because few so-called general geographers do in fact concern themselves with distributions, or locations in real places.

In the case of a single phenomenon the geographic approach should be to study, first of all, the distribution, density (where applicable) and location. A more precise contrast with the second approach above emphasises the study of the horizontal arrangements of phenomena. This may involve simply the layout of different features of one topic. To quote obvious examples, it has long been recognised that there is a logical arrangement of the arêtes, cirques, U-shaped valleys, moraines and outwash plains around the peaks in a glaciated landscape, or of houses, factories, communications and public utilities around the central business district of a city.

The study may be confined to one topic such as landforms and urban land use, or involve several topics, but in general geography all the phenomena listed in Chapter 2 are not combined in one study. However, the location of, say, settlements in relation to rocks, relief, soils, land use and communications, or industry in relation to labour, markets, fuel and raw materials can involve several different types of phenomena. Possibly the functional regional approach should be included here, for there is the implication that these phenomena are not arranged horizontally in space in the manner observed just by chance, but that all the parts have some functional connection. This is obviously so if one sets out to delimit functional regions, but is equally true in the case of location of industry, urban zones, service areas, landforms, and the position of zonal soils in relation to vegetation and climate. The danger here seems to lie in the possibility of the geographer being misled into studying the functioning of these systems; into studying the systems themselves, rather than concentrating on the spatial relationships of the parts. Moreover, from a geographer, one would expect particular attention to the arrangement in the landscape in real places on the earth's surface. However, the realisation that he is studying certain aspects of systems should help to direct his attention to causal connections and relationships rather than coincidence and mere juxtaposition. A factory may be located in relation to raw materials some distance in one direction and a market some distance in another direction, and there may be no connection with the firm next door.

There is one other method which has also been divorced from real places on the earth's surface. Many years ago Christaller[14] needed to imagine an isotropic surface in order to explain the location, spacing and sizes of central places. Spatial analysis[15] is now developing rapidly, and its relevance to geography is clear enough, but too often when reading articles in this expanding field, one gets the impression that the writer finds the relief, shape of the continents, street plans of actual towns etc. most disruptive factors both in the spatial arrangement of the topic in question, and to the neat theories, models or formulae which could be formulated if only the Rockies, Atlantic and medieval towns did not exist. One can imagine a mathematician being so upset, but it is difficult to believe that these analysts are in fact geographers.

These five points are aspects of the same basic approach, that of studying the spatial arrangements of phenomena on the earth's surface. The phenomena may be symbolised by dots, lines, surfaces or by three-dimensional models[16] and the main concern is with their distributions, patterns, networks, connections and interactions on that surface. Chapman has formalised this approach into the study of five 'conceptual elements': location (e.g. site, situation, place), spatial distribution (e.g. continuity, discreteness, pattern, density), spatial association (e.g. interdependence, coincidence, complementarity), spatial interaction, and spatial systems.

In spite of the similarity of wording one doubts whether the study of spatial systems, as systems, can be classified with the study of the location, distribution, association and interaction of phenomena on the earth's surface. But a more important point here is that while one welcomes the fact that Chapman has put this emphasis on space, on the fact that the geographic approach is to study relationships in space, one would emphasise that this refers to a particular part of real space. This is not the geometrical space of the mathematician, just as it is so obviously not the space of the astronomer. When it is necessary for a worker to ignore the qualities and irregularites of the earth's surface completely, then his work parts company with geography.

1. Lukermann, F., 'Geography as a formal intellectual discipline, and the way it contributes to human knowledge', *The Canadian Geographer,* vol. VIII, no. 4, 1964, p. 167
2. Hartshorne, R., *Perspective on the Nature of Geography,* Murray, 1959, p. 178
3. Hart, J. F., *The Southeastern United States,* Van Nostrand, 1967, ch. 8
4. Hartshorne, op. cit., 1959, p. 181
5. Brunhes, J., *Human Geography,* Harrap, 1956, p. 23
6. Chapman, J. D., 'The Status of Geography', *The Canadian Geographer,* vol. X, no. 3, 1966, p. 133
7. Hartshorne, R., 'The Nature of Geography', *AAAG,* Lancaster, Pennsylvania, 1939, p. 120
8. See Minshull, R. M., *Regional Geography: Theory and Practice,* HUL, 1967
9. Humboldt, *Kosmos;* Ritter, *Erdkunde;* Ratzel, *Anthropogeographie*
10. Turnock, D., 'The Region in Modern Geography', *Geography,* vol. LII, 1967, p. 374
11. Minshull, op. cit., chs. 1 and 2
12. Herbertson, A. J., 'The Major Natural Regions', *Geographical Journal,* 25, 1905
13. Whittlesey, D., 'Southern Rhodesia, an African Compage', *AAAG,* vol. 46, 1956
14. Christaller, W., *Die Zentralen Orte in Suddeutschland,* Jena, 1933
15. Berry, B. J. L., and Marble, D. F., *Spatial Analysis,* Prentice Hall, 1968
16. Cole, J. P., and King, C. A. M., *Quantitative Geography,* John Wiley, 1968

5

PRACTICAL APPROACHES

Given the general idea from the last chapter that geography is the study of the arrangement of certain phenomena on the earth's surface, one needs to consider the approach of individual geographers in practice. In the persons of Varenius, Humboldt, Ritter and Ratzel we have examples of men who attempted the full geographic task of studying all the relevant topics throughout the world.[1] Both the amount of information and the intensity of detail demanded nowadays make it impossible for one man to produce an original world geography. Even simplified world geographies based on secondary sources such as those of Stamp and Unstead seem to be things of the past.[2]

The practical problem of the geographic approach, then, is for one geographer to choose less than the entire surface of the earth, or less than the total number of phenomena considered in Chapter 2. Both practical approaches have been used, and each has been in vogue at different times. Varenius, in his *Geographia Generalis* published in Amsterdam in 1650, attempted both, and gave the names which we use today. The plan of his unfinished work included general or universal geography, and special or particular geography. The words general and special are still used, but much more often we use the incorrect word 'regional' when we mean special geography. After Varenius the two approaches were often combined, in the sense that a general geographer, studying a phenomenon throughout the world, would divide it into regions for ease of handling. However, in the work of Humboldt, Ritter and Ratzel the emphasis was on general geography, until the pendulum swung in the other direction with the work of Hettner (1895–1942) in Germany and De la Blache in France who put the emphasis on regional geography. At present the pendulum is swinging to general geography again, some think for the last time.[3]

From the chapters on content and method we have seen that geography deals with some sixteen topics as they vary from place to place on the earth's surface. Geography as a whole studies every variation of each topic in every part of the world, and then attempts to describe and explain these different variations and combinations. As the explanation often takes more time in research and thought than the observation and description, obviously the task of producing a complete regional geography of the world is beyond one man. However, not only must the work be divided up, but different geographers have completely different interests, motives and methods within this total framework, and tackle different parts of the work. From a purely theoretical, idealist point of view, the only snag is that there is not a conscious working toward a complete world geography, because there is not, and should not be, an overall direction and control from one point. Thus much work is repeated. Germany, France, Britain and the USA have all produced repetitive, overlapping works on continents such as South America. In contrast, other regions and topics are often ignored by everybody; for example, until recently the only book on Africa was thirty years out of date, and good works on the geographical distribution of vegetation and soils are still few and far between. Since the days of the complete geographies of Varenius, Humboldt and Ritter, possibly when the amount of data was much smaller, the main attempt at a complete work has been the French *Géographie Universelle* published by Armand Colin in Paris.

Bearing in mind that each topic of geography, in the abstract, is not at all important to geographers by itself, we see two ways in which geography is the study of the interrelation of phenomena on the earth's surface. First, there is special geography. This has been called regional geography so widely for so long, that we tend to forget that a general geography such as climatology is also regional when it describes the climate in each climatic region of the world. However, using the words region and regional in the commonly accepted senses, regional geography studies a part of the earth's surface in depth.

The size of the area of study is usually a continent or less; down, sometimes, to the size of a small English parish. Within this area or region the geographer may study as little as two or as many as sixteen basic topics, and may then have to refer to a much wider number and variety of topics in order to explain what he has found. Within the region the work may be purely systematic, the geographer working steadily and thoroughly from rocks to population, or it may be a compage, the geographer selecting his topics and rearranging their order according to what he finds in the region.[4]

Regional geography has been under attack from several directions[5] but the author's conclusions in *Regional Geography: Theory and Practice* are that this is one of the many swings of the pendulum which geography has endured,[6] and that regional geography is such a large and essential part of the whole that it must flourish, but probably in modified form. The attacks have really been on superficial features such as the regional geographer's inability to draw precise boundaries, and his reluctance to use the new mathematical methods. Regional geographers, however, are interested in the cores and characteristics of the regions, and precise boundaries do not exist and do not matter. Whittlesey's[7] compage method turns the so-called defects of the regional method into advantages and makes them essential features of regional geography.

In addition to studying one area in depth, which at the extreme is the study of the sixteen-topic total-region or compage, regional geography compares and contrasts one region with one or more others. Even to the most casual observer some parts of the country and of the world are obviously different from others. The regional geographer studies, describes and explains these differences in a detailed, disciplined and systematic way, so that traveller's tales and wonders of the world become an orderly body of knowledge which gains immense value from the comparisons, contrasts and causal connections which are thus revealed.

There has been much argument about whether regions exist, or are just a convenient idea for the geographer. What does matter is that regional differences exist and the regional geographer studies these. Sometimes a region, such as the English Fenland, is self-evident and can be mapped precisely; at other times the geographer has to make artificial divisions to facilitate his work, for example to divide up a huge area like the African plateaux.

Regions may vary greatly in size, detail and number of topics, but the basic division is into formal and functional regions. The Americans call these uniform and nodal regions respectively, which is more descriptive. The formal or uniform region is more or less similar throughout its extent *in the topics under discussion only*. A relief region may be uniformly level, a climatic region uniform in rainfall and temperature, a farming region may be uniform in rock, relief, climate, soil and farming, within specified limits of deviation, but is not necessarily uniform in any other topic.

The functional region is diverse, and often has its functions centred on a town or city, hence the name nodal. The functional region may be a small area round a market town where the surrounding countryside produces a variety of furs, timber, fish, minerals and food while the town in turn produces clothes, furniture and tools, and acts as the centre for transport within the region and communications with other regions. At the other end of the scale a huge area like the USSR may be so varied, and produce all of the commodities required by man, that it is basically self-sufficient, and, given adequate internal communications, can be regarded as a closed system.[8] Regions are classified by: type (formal or functional), rank (see table 1, column A), and category (the number of topics—see table 1).

An examination of general geography does reveal some difference between theory and practice. If, in theory, regional geography examines many topics in a few places, then general geography should examine one or two topics throughout the world. In practice, one finds that many self-styled general geographies deal not with the areal distribution of a phenomenon throughout the world, but with the principles of the nature of that phenomenon.

One attitude to this state of affairs would be to accept this as the nature of geography, that in reality general geography is largely the principles of geomorphology, principles of economic geography, principles of town sites, plans and functions, and so on. But the disturbing fact, which forces one into the attitude of saying what a branch of geography *ought* to be, is that often these works appear to be general conclusions based on an examination of the world distribution of the phenomenon in question.

The sixth-former, and the college student in particular, are in

TABLE 1 : CATEGORY

Rank A.	Example B.	Single topic	Combined topic	Multiple topic	Total region	Compage
CONTINENT	N. AMERICA	VEGETATION REGIONS	CLIMATE REGIONS i.e. TEMPERATURE RAINFALL PRESSURE WINDS	CLIMATE SOIL VEGETATION	ROCK RELIEF and DRAINAGE CLIMATE SOILS VEGETATION TRAPPING FISHING FORESTRY FARMING MINING MANUFACTURING COMMUNICATIONS SETTLEMENT POPULATION POLITICAL UNITS	Varying number of topics in best order to describe that particular region
MAJOR DIV. or REALM	WESTERN CORIDILLERAS					
PROVINCE	SIERRA NEVADA					
SECTION or DISTRICT	LINE OF VALLEYS					
TRACT, PAYS or LOCALITY	GLACIATED VALLEY					
STOW	ADRET or UBAC					
SITE	A FLAT or A SLOPE					

The ranks given here are compiled from work by Fenneman, Linton, Unstead and Whittlesey, hence two or more names in some cases.

danger of accepting these principles as valid conclusions when too often they are unproved theories. The author has insisted (sample and example) that detailed and exhaustive examination of phenomena must come before any valid conclusions (see p. 80). Not only do many works of general geography fail to present their objective facts first, but there is often no evidence that sufficient research has been done first. The result is the situation familiar to many practising geographers and to teachers who have to be intellectually honest with their pupils in the field, that in real life there seem to be more exceptions to, than examples of, some of the rules stated as law in the books. This leads to less able teachers ignoring the bulk of evidence in the field, and taking their long-suffering students on long coach journeys to see the one case which proves the rule.

Thus on two counts we can say what general geography ought to be. On the count of its theoretical position in the scheme of things and on the count of the practical need of much more detailed research. Another aspect of the work all too familiar to those in colleges and schools alike is the simple absence of factual information about topics both here and abroad. In theory, then, general geography studies the whole world, but only in respect of one or two topics at a time. The general geographic study of population is a good example of a world study restricted to one topic. The density, distribution and composition of the population vary infinitely throughout the world from areas with no people at all to places like Java and London with over 3000 per square mile. Moreover the constitution varies from young populations with, say, many factory wage earners, through balanced populations in farming areas to populations of elderly people in Hove and Cheltenham.

The world population map in any atlas will illustrate the topic which the general geographer studies. In order to deal with the distribution he may divide it into single-topic regions of dense, medium and sparse population. In order to describe it he will use graphs and age-sex diagrams, and in order to explain it fully he will have to refer to every other topic of geography and many topics completely outside, or marginal to, geography such as nutrition, technology, education, health, religion, birth-control, politics, immigration policies, history and folklore.

The general geographer, in this ideal sense, examines things as they actually are on the earth's surface now; and draws conclusions and suggests laws later. As many eminent geographers have found after a lifetime's experience, there are so many factors affecting real-life distributions and variations, and the possible number of permutations is so great, that very often they refrain from formulating laws and insist that the study of geography in the end comes down to a study of special cases. This is a most encouraging conclusion, for it suggests that human life is infinitely variable and interesting, and that man is not just an animal bound by instinct and physical laws. The general geographers who lean toward natural science seem to be unhappy about the unpredictable nature of man, and insult their own kind by trying to see him as a predictable clockwork toy.

There may be ranks of general geography, in the amount of detail in different works, and there are certainly categories of general geography in that it may be single-topic, combined-topic or multiple-topic. Physical works such as climatology tend to be combined-topics while Human works such as economic geographies, or social geographies dealing with rural settlement, house types, field patterns, roads, towns and population, tend to be multiple-topic works. Here we see the gaps in the work done, and the trend to principles rather than description. There is not really a general geography of relief[9] and even the atlas of landforms of the United States[10] gives only selected examples. It is suggested, as a theory to be proved, that were the landforms of North America, Britain and Europe examined in detail, periglacial conditions would oust river action and glaciation proper as the main factor in the explanation of the landscapes in which the majority of active geographers this century happen to live and work. Similarly, we need much more general geographic study of such topics as communications, field systems, land tenure, rural settlement, towns and cities to fill large gaps in our factual knowledge. Too often generalisations in British geography are made from a few samples in southern England. Too often generalisations are made about the world based *only* on research in Europe and North America. The greatest contradiction in general geography is its preoccupation with principles which tend to gloss over areal differentiation.

At best, regional geography and general geography are complementary, and are the two parts of geography necessary to keep the balance. Regional geography examines areas in detail, examines special cases, tries to correlate a great number of topics, but cannot make general laws. General geography examines the world in one feature at a time and thus has a wider view, can analyse the factors affecting one topic at a time, and can make general laws. The two are complementary in that regional geography can provide the wealth of factual information for the general geographer (he can't visit every region himself) while general geography provides the framework of laws and principles within which the regional geographer's work has meaning.

The main decision, then, in the practical approach of one geographer to one particular piece of work, is whether to select an area or a topic for study. Some people consider themselves regional or general geographers and have made this decision already; or the particular problem for study will determine the nature of the approach. Berry[11] shows that these approaches are just two extremes of a continuum. At one extreme is one region in which one studies the arrangement of all the phenomena. At the other extreme one studies the geography of one topic throughout the world. In between are infinite variations of the combination of many topics in a few regions, or one or two topics over a large part of the world. Thus a necessary process in the main decision-making is to decide both the number of topics and the size of the area at the same time.

While the basic problem of defining the amount of work to be tackled remains, other problems facing the geographer have changed. In the second half of the nineteenth century and well into the twentieth there was a serious debate as to whether human phenomena formed part of the proper content of geography. With the rapid advances in geology, geomorphology and climatology, what we now call physical geography became Geography in the eyes of many authorities. We still have the irksome legacy of this attitude in that some geographers and teachers still call rocks, relief, climate, soil and vegetation 'geographic features' and 'geographic factors' and, much worse, still think in terms only of physical stimulus and human response.

The geographer no longer needs to ponder whether he should study human phenomena just as thoroughly as physical phenomena, but increasingly he will be faced with a question just as ridiculous. One can foresee the time when a geographer who thinks in terms of the landscape and the real earth's surface will be considered very old fashioned. With the increasing interest in horizontal interaction, in systems, in spatial relationships, and the use of models and mathematics to analyse them, the landscape is becoming just a nuisance to some new geographers. Many of the hypotheses, simple models and even complex formulae will only apply to a flat, featureless surface. Time and again an isotropic surface is postulated at the beginning of the work. Position and distance are the things in vogue at the moment, just as rocks and landforms were in vogue in W. M. Davis's day.

In contemporary studies of agriculture, distance from the farm and from markets are much more important factors than soil and climate. Studies both of the distribution of towns and the internal structure of towns put the emphasis on spacing, distance apart or from the centre, cost of transport, cost of land, rather than on the relief and shape of the country, or on historical development. Clearly position and distance are factors as important as rock and relief, factors which have had less attention in the past. Clearly, the new theories, models and formulae must be worked out under ideal conditions, but anyone who is aware of the one-sidedness of geography in the past must be aware that these ideal generalisations about spatial relationships can too easily become mistaken for statements about reality itself.

One wonders whether the people who derided the idea of formal regions—an impossibility of uniform rock, relief, climate and soil—are those who need to postulate a flat, homogeneous plain with uniform climate and soils for the development of their land-use zones, industrial networks, transport networks, central places, and so on. Not only do the physical features vary greatly, and interfere with human activity, but also one type of human activity interferes with another, and distorts any ideal pattern which might be imagined.

Physical geographers are aware of the effect of the real landscape on the features they study. Far from ignoring the landscape and playing down its effects, they take full account of it. Thus the effects of rocks on relief, of relief on climate, of rock, relief, climate and vegetation on soil, are studied thoroughly, not played down. Possibly this need to play down or ignore the effects of limestone, mountains, wide rivers, forests and coasts in the spatial organisation of economic activities, transport and settlement simply demonstrates that such general human geography is at a very primitive stage in its development. But those enthusiastic about spatial analysis are ignoring factors in the real landscape of which geographers have been aware for centuries.

This aspect of geography is changing rapidly, and one can only hope for one of two things. Either, that spatial analysis develops as a discipline separate and different from geography (but with its relevance to geography) as has happened to geomorphology. Or, that both the researchers in this field, and those students who read their works, are perfectly clear that these geometrically neat ideas, undistorted by any landscape, are only ideals with which to compare reality. Geographers will then be saved from further confusion, and can use these spatial theories to help them further understand the real landscape, varied and irregular as it may be.

Finally, two other minor ways in which the size of the geographer's task may be limited must be mentioned. Obviously the amount of detail will affect the length as well as the accuracy of the work. Haggett[12] considers this point, but also goes into some detail about how the scale of the work can affect the type of explanation which is called for in geography. He maintains that less detail, on a wider scale, calls not only for more generalisation, but also for a different type of interpretation. He also suggests quantitative techniques to help to determine the different levels. Less obviously, the arrangement of the topics in the study, and the order in which they are tackled, will make a difference to the amount of work. Geographers attempting a systematic approach like that of the old style regional geographers must necessarily plough through rock, relief, climate and so on to the inevitable result of producing a brochure, reference book or encyclopaedia. By selecting any one topic or problem for analysis, one immediately avoids this problem, providing one then considers only those other phenomena relevant to the theme or problem, and avoids including all the other details of the geography just because they are there. Whittlesey called this the compage approach.[13] With some amusement the author has seen Whittlesey's word scorned, by people who have used the approach under other more pretentious names.[14]

The practical approach necessitates that one geographer attempts less than a complete geography of the world. In practice we see that his task is limited either to some part of the world, or to a few phenomena. Moreover, as Hartshorne, Berry and others show, this is perfectly acceptable on theoretical grounds. One result of this necessary division of the work between geographers, however, is the familiar dichotomy in geography: general and regional, systematic and compage, physical and human, landscape and spatial, single topic and multiple topic. The oldest and most familiar is the contrast between general and regional, followed in the nineteenth century by the split into physical and human. These are necessary to enable individuals to attempt a piece of work within their capabilities, according to their interests and preferences, but they have led to arguments so bitter that geography has been in danger of splitting into several contrasting disciplines. The author would insist that on both theoretical and practical grounds this dichotomy is necessary, because the task of geography is beyond one man. What one must deplore most strongly is that geographers in any given speciality act so childishly as to think their speciality comprises the only worthwhile geography, and to vilify workers in other specialities. One cannot stress strongly enough that while one man may well have to be a regional or general geographer, specialise in physical or human topics, geography is a unitary discipline which must embrace the study of all types of phenomena throughout the world.

1. For the history of geography see:
   Dickinson, R. E., and Howarth, O. J. R., *The Making of Geography*, Oxford, 1932; Hartshorne, R., 'The Nature of Geography', *AAAG*, Lancaster, Pennsylvania, 1939; Freeman, T. W., *A Hundred Years of Geography*, Duckworth, 1961; Houston, J. M., *A Social Geography of Europe*, Duckworth, 1953; Taylor, G., *Geography in the Twentieth Century*, Methuen, 1957; Dickinson, R. E., *The Makers of Modern Geography*, Routledge and Kegan Paul, 1969.
2. Stamp, L. D., *The World*, Longmans, 1950; Unstead, J. F., *A World Survey*, ULP, 1955; Willis, M. S., *A Systematic Geography of World Relations*, Philip, 1950; James, P. E., *A Geography of Man*, Ginn, 1959
3. Saey, P., 'A new orientation of geography', *Bulletin de la Société Belge d'Etudes Géographiques*, tome 37, no. 1, 1968
4. Minshull, R. M., *Regional Geography: Theory and Practice*, HUL, 1967, ch. 9
5. Haggett, P., in Chorley, R. J., and Haggett, P., *Frontiers in Geographical Teaching*, Methuen, 1967, p. 15; Kimble, G. H. T., in *London Essays in Geography*, Eds. Stamp and Wooldridge, Longmans, 1952, ch. 9.
6. Freeman, T. W., *A Hundred Years of Geography*, Duckworth, 1961, ch. 1
7. Whittlesey, D., in *American Geography: Inventory and Prospect*, Eds. James, P. E., and Jones, C. F., Syracuse University Press, New York, 1954, p. 19
8. Hooson, D., *The Soviet Union*, ULP, 1966
9. Taylor, G., *Geography in the Twentieth Century*, 1957, p. 113, reference to Fritz Machatschek, *Relief of the World*
10. Scovel, O'Brien, McCormack and Chapman, *US Atlas of Landforms*, US Military Academy, West Point, New York (John Wiley), 1965

11. Berry, B. J. L., 'Approaches to Regional Analysis', *AAAG,* vol. 54, 1964, p. 2

12. Haggett, op. cit., p. 4

13. Whittlesey, op. cit.

14. Grigg, D., 'The Logic of Regional Systems', *AAAG,* vol. 55, 1965, p. 1; Lewis, G. M., 'A new approach to regional geography', *TIBG,* no. 45, September 1968, p. 11; McDonald, J. R., 'The region, its conception, design and limitations,' *AAAG,* vol. 56, 1966, p. 517; Turnock, D., 'The Region in Modern Geography', *Geography,* vol. LII, 1967, p. 374

6

CHANGING TECHNIQUES

No claim will be made that any techniques are unique to geography, but an examination of those which are used most often by geographers will help to throw some light on the nature of the discipline. Moreover, as techniques change or new ones are employed one needs to consider whether they in turn tend to change the content, approach and purpose of geography.

The oldest technique is the art of literary description. Geographic description is in disrepute in some quarters at the moment, partly because much of it has been mediocre and inaccurate, and partly because we are in a phase when even accurate description is considered not to be a sufficient end in itself. While geographers are not unique in this, they have a strong tendency to ignore all old techniques when a new one is adopted. Like children they abandon all their old toys when presented with a new one, instead of refining and perfecting their equipment by rejecting only what is worn out and useless, preserving what is of value, however old, and accepting only those new techniques which will help to make real progress.

Whittlesey wrote: 'The presentation of the results of regional study has its frontier no less than the methods of investigation. An understanding of the elementary principles of English composition, of the sort so often neglected in our schools and colleges, is fundamental. This involves the construction of a properly balanced outline of major and minor headings, an appreciation of the function of the paragraph, role of the topic sentence, and a clear view of the content of an effective introduction and conclusion. Training is needed here, not experimentation, for the basic principles have long been known.'[1]

Darby[2] agrees with this view, but believes that good geographical description is one of the most difficult techniques to master. In elaborating on some of the pitfalls, he also gives some most valuable advice on how to make description as accurate and concise as possible. The poor examples of description have usually been combined with those works which attempted to give a complete account of a region, country or continent. When there has been neither geographic approach nor any purpose beyond description, then these catalogues of a country's contents have been mediocre indeed. Yet precise description still remains one vital part of geographic work, description using other media as well as words, and good English is still one of the necessary techniques.

At the other extreme from the presentation of findings and conclusions lies the main technique for collecting information and making observations, namely fieldwork. In the author's experience few other words besides fieldwork have so many different meanings to different people, and so some clarification is offered here. First, one would insist that the geographer goes out to observe and record, and not to gather proof for his prejudices. Thus one takes exception to statements which say that geographers go out: 'to show that relief is the result of physical processes',[3] or 'to show how environment has influenced man'.[4]

In teaching, and in training future geographers, fieldwork includes several kinds of activity.

(1) *The excursion* to give first-hand experience of phenomena, landscapes and distributions.

(2) *Practice in techniques*. Many professionals insist that training in the techniques of observing and recording is given in well-known areas on well-known phenomena before original work is done in new areas.[5]

(3) *Original fieldwork* in which the student uses the basic techniques of:

Recording data on a base map
Making an original map
Counting such things as people or traffic
Completing questionnaires
Field sketching

to bring back data necessary to his analysis.

(4) *Collecting documentary evidence in the field.* In local collec-
tions, museums, offices of local authorities is to be found both
original source material and secondary documentary material.
Either because this information is not available through any
agency, or because it avoids repetition of work already done, it
should be collected while 'on location'. The dangers of local
documentary material collected during the course of fieldwork are

(a) It may be inaccurate, incomplete or out of date.
(b) Much tends to be of use to historians or economists rather
    than geographers.
(c) It encourages some to become armchair geographers
    completely.[6]

(5) *Testing an hypothesis* in the field. (6) *Field teaching* is
mentioned here as distinct from point (2) and as something to be
done very carefully to avoid 'reading into' the landscape things
which cannot be directly observed on the ground and which
have been learned from other sources. So one would stress
'work' in the sense of making observations and recording them
accurately; and at times it seems necessary to explain 'field'.

To many laymen, a geographer is someone who spends his
time in tweeds and Veldtschoen hammering rocks in Cumberland,
or in anorak and wellingtons lost in Wales. When I start to enthuse
about the Lake District, so many people say 'Yes, it's such a good
place for geography'—as if there was more to learn there than in
Market Street in Manchester or Waters Green in Macclesfield.
So two common misconceptions must be put right at the start.
First, the geographer spends only part of his time in the field.
Second, fieldwork is just as important, and more exacting, in an
area of intensive farming or in the centre of a big city as it is at
Spurn Head or up the Cuillins.

In fieldwork one geographer can cover only a very small area if
he is to know it in detail. To drive this point home, imagine that
one is fit, and can walk fifteen miles in the country in one day,
recording observations as one goes. Draw a route fifteen miles
long, coming back to the starting point, on a one-inch ordnance
map. Then see how little of just one sheet will have been walked
over once. So one geographer, relying only on his own fieldwork,
can either cover a tiny area in detail, and write a total-regional

account, or cover a large area and write in less detail or confine his attention to certain topics of general geography.

Obviously, few geographers limit themselves to the results of their own fieldwork all the time. The possibilities then are either to combine one's own observations with material from maps and documentary sources, or to combine the fieldwork done by several other people. Usually this latter method consists of using degree theses and papers published in the journals, but sometimes the writer may have several people doing field research for him. For example, it is difficult to imagine one person having the time, money and energy to study the geography of even one topic in the field throughout the world. Thus at one extreme, complete field-work is impossible, but at the other extreme a certain amount of fieldwork is essential for the geographer to 'sample' the area or topic of study. In order to produce a major work such as the geography of North America or a world economic geography the geographer must use the results of other peoples' work and combine them; but also, he must investigate some things at first hand, in order both to check the validity of his secondary sources and to be familiar personally with what he describes. For one person to write about a large area or several topics there must be a combination of sample fieldwork and the use of documentary material.

The ideal situation would be to have many fieldworkers pro-ducing the original observations for one writer to analyse and use. In fact there are too few people doing original fieldwork and quite often their findings are either never published or have to wait several years before they can appear in the journals. So it is a commonplace for the geographer to have to make use of data produced for some other purpose and not presented in the form most useful to him.

Whatever the source of the data, however, the geographer must spend much of his time in the library, the map room and his study. More important than the fact that to some people geography seems to cut across other disciplines, is the fact that the major geographi-cal work takes the results of work done by other types of experts and by other geographers. The maps which are the end product of the geologist, surveyor, pedologist, ecologist, and (in some coun-tries) land-use surveyor; the figures which are the end products of the weather stations, Board of Trade, Ministry of Agriculture, Ministry of Transport and the Registrar General; the detailed

studies by geographers of some small area or specific topic—
these are all the raw material of the regional or general geography.
This raw material in maps, tables, pamphlets, reports and journals
has all to be found, sifted, read, digested and then organised into a
finished, logical book. The geographer cannot be roaming the
Highlands or the Savanna all the time; he must devote consider-
able time to reading, thinking, re-thinking, writing and re-writing.

Fieldwork in research consists, then, of three types: full
original work on a small area or a limited topic, sampling a large
area or a multiple-topic study, filling gaps not covered by geologists,
surveyors, census-takers etc. in some parts of the world. The
geographer must be able to do this, but he does it only when
absolutely necessary, for two good reasons: there is no value,
in research, of doing work already done and if the geographer
had to do the fieldwork of geologist, meteorologist, demographer
etc. he would never have time for geography.

If we distinguish sources of information which the geographer
may use from the techniques he employs to examine those
sources, then the prime source is the earth's surface, and the major
techniques are fieldwork and the study of maps and air photographs. As suggested above, the maps and vertical air photographs are the media by which the observations of thousands of
people can be made available for study by one person.

The maps in question may be made by other geographers or
specialists in other disciplines, but more commonly they are made
by surveyors and cartographers for general use. Thus one of the
basic tools of the geographer is not made specifically for him, and
the limitations of the map must be understood. When one
compares an ordnance survey one-inch map with a monochrome
air photograph, or even with the vividly coloured landscape from
a low-flying aircraft, one at once realises the distortion of showing
as thick, bright red lines the roads which are insignificant from the
air. This obvious example reveals the different purposes of the geographer and cartographer. So many maps for general use
emphasise the roads and towns because their main purpose is to
show the layman how to get from one place to another, or a
rocket crew how to aim at the enemy target.[7] But the geographer
tries to use these maps to study distributions, correlations, and
spatial relationships of many other phenomena. Clearly, topographic maps are not sufficient for this, and the word map
includes geological, meteorological, climate, soil, vegetation, land

use and population maps which are not as familiar to the layman.

Most of these maps can be drawn directly in the field, and provide data similar to that obtainable from vertical photographs, but climate and population are phenomena of a different kind. In contrast to visual observation on the ground, the data is in the form of readings of temperatures, rainfall, pressure and so on over a long period, and of a record of all the people made on one day. The resulting maps show something which cannot be seen on the ground, and climate and population are the two phenomena which cannot be studied in air photographs.

Map language

In recent years there has been discussion among planning authorities and conservation societies about methods of recording and classifying landscapes. This in itself is most encouraging, because it shows that other specialists are becoming aware of the finer gradations of landscape (say, within one county) and are more concerned to preserve them and to dovetail in developments more carefully. This trend makes even more incredible the fact that they are trying to record and compare landscapes by means of a card index system.

While these authorities have not the money and the advanced equipment of the Agricultural Stabilisation Service of the USA, which has a complete cover of vertical air photographs of the country, they obviously have enough money and equipment to think of using computers. Some of these authorities have turned to geographers for advice on how to make out these card indexes, presumably imagining something like the admirable system which Brunskill has worked out for recording houses.[8] Although British geographers might welcome a complete air photo cover, they are more than satisfied with the superb map coverage, especially in the 1:63,368 and 1:25,000 sheets, and point to this as the answer to the planners' dreams. The position at the moment seems to be that the planners blame the geographers for not helping them to work out a system, while the geographers do not understand how anyone can wish for anything better than our topographical maps as records and indexes of the landscape.

This misguided aim of the planning authorities, and the misunderstanding between the two groups of people, arises from lack of fluency in some of the skills under discussion. If the planners want punched cards and tapes to run through computers, when the

C

maps exist already, then one can conclude only that they are so well trained in mathematics that they try to use this medium for everything, and so poorly trained in map reading that they fail to appreciate the perfect tool at their disposal for about £50.[9] Basically, this failure in the skill of map reading is the fault of some geography teachers.

In another sphere of activity, a historian quickly appreciated the value of British 1 in maps and at one stage was strongly recommending them to the geographer. Professor Hoskins, who was a lecturer in economic history and became professor of local history at Leicester, not only repeatedly stressed the value of topographic maps in presenting a wealth of detail in the best possible manner, but also made practical use of them in his researches.[10] The best way to be convinced of the value of one-inch ordnance survey maps in showing the different types of landscape is to lay out two or three sheets together on the floor. Sheets 171 and 183 will give a bird's-eye view from central London, across the downs and weald, to the channel, vividly showing the extent of the city, and a surprising amount of woodland. Sheets 112, 113 and 114, in their combination of relief, vegetation, rural settlement and drainage lines, clearly show the landscapes of the coalfield, Trent valley, scarp, fens and wolds.[11]

This use of the wrong medium, say maths instead of maps, is often seen in the arts, when some creative artist fails to express himself well because he is using the wrong medium, film instead of the theatre, prose instead of verse, painting instead of sculpture. This usually results from the fact that one needs long and intensive training in the skills and techniques. A man or woman trained from the age of three in ballet, piano, painting or whatever, may later turn to these familiar techniques to try to express some emotion best expressed in another way. How often have we, as laymen, wished we could paint, write poetry or compose music when some strong feeling has been aroused and have had to make do with prose?

Here is the point; we all have some facility in English and maths, however slight. To advocate training in music, painting, drama and sculpture may well be going too far in view of the time and money available, yet the better schools do it. But there is a more realistic case for insisting that the well-educated person, who may become a town planner without any special training in geography, should be as competent in reading maps and in sketching as he is in writing and calculating. Then, and only then, we can hope that the most appropriate medium will be used for a given job.

In trying to fill some serious gaps in his own education, the author has opened what claim to be elementary introductions to certain subjects, only to find, say, a few bars of music or a pair of equations used as part of the explanation on the first page. If experts in music and mathematics, explaining to the layman, can use their 'languages' of notes and figures to describe, without further translation into English, then the expert geographer certainly can use his map language to describe without any necessary recourse to repetition in words. The well-drawn map is easier for the layman to understand, because it is a picture of reality, rather than the hieroglyphics of abstract thought as is the case in mathematical and musical notation.

Maps are the essential tools of the geographer, and could be used as his language much more, in the sense that his examples and descriptions could be presented to the general public in map language rather than in words. This could not happen suddenly, for it requires training in school, so that later the student can 'read' what the teacher has 'written'. It may seem that it is too much trouble to teach everyone to read maps, and to urge geographers to describe by means of maps, when they both have the written language as means of communication. But any geographer who has ever tried to describe in words a phenomenon, say the distribution of population, which varies at different rates in different directions, and then has plotted the same thing easily by means of dots on a map, will understand. Possibly the reader is familiar with geography texts which give the map and a written commentary, and has been confused when the involved commentary seemed to bear no relation to what was shown on the map.

Because many geographers still feel the need to repeat what they have shown on their maps in a commentary of words and figures, those maps have to be compiled and drawn in such a way that they can be re-described in words and figures. This results in two faults, one more serious than the other. Firstly, it makes mapping much harder. Secondly, and more seriously, it tends to exclude from the maps the rather more nebulous qualitative features which are not capable of being given precise numerical value and being put through a computer. Some maps may have to be pictorial rather than precise plans, but if they show phenomena which can be treated only in this way, then they are the most appropriate medium. Geographers, to some 'jacks-of-all-trades', make use of information from any other discipline. Increasingly they are making use of all techniques of presentation and must continue to do so; but just as the novelist uses words, the mathematician figures, the historian quotations, the scientist graphs and diagrams, the geographer uses maps.

Geographical description has been mentioned primarily as a technique of presentation; fieldwork primarily as a technique of obtaining data. Mathematics will be mentioned as one technique vital in analysing that data. As a final comment on maps and air photographs it is clear that they have the unique distinction of being vital at all three stages of the geographer's work. Maps and vertical photographs contain the data exactly as the geographer wants to study it, as it is distributed on the earth's surface. In his process of analysis and re-synthesis the geographer may work on these maps, or re-plot certain data, making precise measurements and counts and applying formulae to obtain quantitative results about degrees of correlation, relative location, and so on.[12] Finally, other maps will be drawn and photographs taken to illustrate the conclusions made in the geographer's published work.

Anything written in the days after Apollo 11 must look to rapid technical change and consider things once thought impossible. Fieldwork, mapping on the ground the hard way, and even conventional air photographs could become completely obsolete. The satellite systems which even now send back pictures of the weather and the landscape from near space are being improved to send back much more precise data on many phenomena. Surveying for minerals, mapping vegetation and plant disease, resource surveillance and so on are already being carried out from satellites. This remote sensing will be extended to traffic and population counts in time, saving much footwork.

In 1965, Rodoman[13] had a seemingly bizarre concept not only of satellites sending data about these phenomena back to earth, but of huge television screens showing a continuously changing, up-to-the-second map of the geography of, say, the USSR (and probably the USA as well). It is hard to imagine a screen so big, and a moving 'map' so complex, but many screens, each showing one phenomenon, over a limited area, probably do exist. At the time of writing, Rodoman's other idea of then having a computer to receive this constantly changing data, to analyse it, and to put out an up-to-date economic plan for each region of the USSR seems frighteningly feasible. The Americans have the same idea in the concept of resources surveillance, by which they mean not finding resources, but keeping an eye on what we have. Thus the satellites monitor harbour silting, cliff erosion, soil conservation, forest fires and disease, shoals of fish, crop ripening, traffic flow and so on. Rodoman's article is one to be read and considered very carefully.

While geographers have possessed sources of information and techniques of recording and presenting their findings for a long time, only recently have they adopted a precise technique for analysing the data. The statistical, quantitative or mathematical techniques have existed for decades. They have been used by a few geographers and many physicists, climatologists, demographers, sociologists and others as soon as they became available, but in general the majority of research geographers did not adopt them until the 1960s and there will be another time lag during their spread into colleges and schools.

The reader is referred to other books for details of the techniques[14] but the point of them all is that they are enabling geography to change from qualitative to quantitative statements. There is, however, some confusion between what these techniques could do for geography, and what they are doing to it. In the author's experience there are two groups of people using these techniques. There are those who understand them, and use them as a means to the end of analysing some geographic problem. There are also, inevitably, those who jump at something new, adopt it because it is new, and not because it is useful for their purposes. Whether they understand these techniques or not, such people tend to stop one stage too soon, when they have the answer to the equation rather than when they have achieved some further understanding of the geography behind the figures. What may confuse the novice even further is that many of the examples of statistical techniques are applied to the data of climate, river flow, economic production, population and so on. They are put forward as techniques useful to the geographer as distinct from the meteorologist, geomorphologist, economist and demographer, and the novice inevitably will lose sight of distributions and correlations in space, as well as of spatial relationships on the earth's surface. The simplest statistical techniques analyse only lists of data such as annual rainfall and population which change in *time*.

What the geographer needs are techniques to analyse distributions of isolated phenomena, networks such as roads, and areas or surfaces. Thus the quantitative techniques most useful to him fall into four main groups: methods of sampling, network analysis and graph theory, correlation techniques, and trend surface mapping. When these techniques are understood, one still has to remember their purpose and function. Most geographers are still trying to do what they always did. They are still trying to make the same kind of statements about phenomena on the surface of the earth. But forty years ago, with less data, and with few mathematical techniques, their statements had to be largely qualitative.

The word qualitative can be misinterpreted very easily. At worst it can mean that some geographers made completely unfounded statements and came to wrong conclusions. At best it means that they were correct, but could not be supported by accurate factual evidence. Geographers have been accused of only looking at maps and landscapes, of only 'eyeballing' them, instead of measuring and collecting quantitative data. At the extreme, some seem to have been entirely subjective in their analysis, but even those who tried to be objective could indicate only the right *kind* of conclusion, and could be no more precise. At that time, the examination questions which began 'to what extent . . .' were impossible to answer.

One would emphasise here that the quantitative techniques have a much more important function than just backing up geographers' statements with numbers. The techniques have a much more vital function at the beginning of the work in revealing valid lines of enquiry. One key point here is the idea of statistical significance. The old eyeballing methods gave an idea of whether two phenomena covered the same area or not; of whether two sets of figures were similar or not. In the past, as the result of this eyeballing, men may have spent years talking about the relation between two phenomena, postulating cause and effect, when in fact the two had come together randomly, by coincidence, and there was no connection at all.[15]

Nowadays, right at the start, one can apply correlation techniques to see whether there is a relationship which is statistically significant. Or one can test for significant difference, and then usefully look for explanations for this difference. Gregory[16] repeatedly stresses this point, that a close correlation between two phenomena does not prove or explain cause and effect; it simply indicates that the connection is so close, and so unlikely to be a matter of chance, that it is worth while spending time looking for the process of cause and effect, and trying to explain it.

Thus the quantitative techniques can help one to avoid the gross error which has been all too easy in a discipline which studies the complexity and variety of the earth's surface. The phenomena are too irregular, too closely interconnected, too dynamic, too infinitely subtle in their gradations, shapes, distributions and relationships to be studied entirely by staring at coarse, incomplete, static topographic maps. For example, sampling methods enable one to avoid bias and subjectivity in choosing data. Similarly trend-surface mapping can iron out all the irregularities to reveal the basic trend not always apparent on a detailed map. Correlation techniques and locational analysis put one on the right track right from the beginning, by deciding significant connection or difference, by putting like with like, and by showing the strongest links in systems and space relationships. Given this, the knowledge, experience, patience, intelligence and other technical skills of the geographer are more important than ever—to use these techniques properly.

Fieldwork, cartography and mathematical analysis are necessary processes which must be carried out before geography can begin. The amount of geographical work being done has so outstripped the supply of information in the form in which geographers need it, that they have been forced, in places, to become surveyors, cartographers and even computer programmers. There will be few cases where a geographer has the help of professional mathematicians, surveyors, field researchers, computer programmers and cartographers, but the danger of geographers having to do these jobs themselves is that some of them think that these necessary preliminaries are geography itself.

1. Whittlesey, D., in *American Geography: Inventory and Prospect*, Eds. James, P. E., and Jones, C. F., Syracuse University Press, New York, 1954, p. 19. See also Rowe, A. W., *The Education of the Average Child*, Harrap, 1959, ch. 6, p. 110 ('books, after all, are still the chief repositories of knowledge and experience')
2. Darby, H. C., 'The Problems of Geographical Description', *TIBG*, no. 30, 1962, p. 1
3. Briault, E. W. H., and Shave, D. W., *Geography in and out of school*, Harrap, 1960, p. 27
4. *UNESCO source book for geography teaching*, Longmans, 1965, p. 2

5. Stone, K. W., 'A guide to the interpretation and analysis of aerial photos', *AAAG*, vol. 54, 1964, p. 318
6. Thatcher, W. S., *Economic Geography*, Teach Yourself Geography, EUP, ch. 3
7. The word Ordnance survey is still appropriate.
8. Brunskill, R. W., 'A systematic procedure for recording English vernacular architecture', *Transactions of the Ancient Monuments Society*, vol. 13, 1965–66
9. The price of a set of one-inch maps
10. Hoskins, W. G., *The Making of the English Landscape*, Hodder & Stoughton, 1955
11. Montague, C. E., *The Right Place*, Chatto and Windus ('When the map is in tune'), 1924, p. 37
12. See, e.g., Haggett, P., *Locational Analysis in Human Geography*, Arnold, 1965; Gregory, S., *Statistical Methods and the Geographer*, Longmans, 1968; Cole, J. P., and King, C. A. M., *Quantitative Geography*, John Wiley, 1968
13. Rodoman, B. B., 'The Logical and Cartographic Forms of Regionalisation and their Study Objectives', *Soviet Geography*, vol. VI, no. 9, November 1965, p. 3
14. See Haggett, Gregory, Cole, above, and also Berry, B. L. J., and Marble, D. F., *Spatial Analysis*, Prentice Hall, 1968; Toyne, P. and Newby, P. T., *Techniques in Human Geography*, Macmillan, 1971; Theakstone, W. H., and Harrison, C., *The Analysis of Geographical Data*, Heinemann, 1970; Yeates, M. H., *An Introduction to Quantitative Analysis in Economic Geography*, McGraw Hill, 1968.
15. Whittlesey, op. cit., warned about this long ago
16. Gregory, op. cit., p. 206

7

CONNECTION, SAMPLES AND MODELS

Other techniques set themselves apart because they are less mechanical and obvious than those mentioned in the last chapter; and because it is genuinely difficult to distinguish whether they are means to the end, or the whole end purpose, of some geographical work. The techniques, mental processes, call them what you will, in question here are such things as taking examples, classifying phenomena, inductive reasoning and making generalisations. These are necessary processes, which people in all disciplines have to use. Yet the means of procedure for geographers are not set out clearly, as are the procedures for fieldwork or a chi-squared correlation calculation. Moreover, some people have completed their work when they have chosen and described certain examples, or when they have made a classification, while others use these processes as steps to a different and more satisfying kind of end product.

Learning such techniques is a different process from learning fieldwork, mathematical or cartographic techniques. No people, no books on logic can tell one how to think. All that works on logic, classification, etc., can do is show us how to check rigorously the interval structure of a chain of reasoning to discover faults and fallacies. Even then, the logic of a piece of work can be perfect, but the conclusion wrong because the premise was wrong. So one has to learn by the long and painful process of experience, and the quickest ways to greater accuracy (which seems to be the aim of geographers now) are this rigorous testing of one's own logic and a continuous effort to learn from mistakes. Three of these vital but ill-defined techniques will be examined here, partly to point out some of the obvious pitfalls, but mainly for the light they throw on the changing nature of geography.

Causality, connection and coincidence

If the geographer's approach to the phenomena on the earth's surface is to study their distributions, correlations, locations and relationships, then geographers have read too much into the landscape and have found causal connections and spatial relationships where none exist.

In the last chapter it was suggested that the maps of the distributions of each phenomenon are compared to find causal connections between two or more phenomena. Thus a comparison of maps of the solid geology, relief, rainfall, soils and crops of a large area might show the same basic patterns in the distribution of wheat and of certain glacial deposits, but completely different patterns of geology and relief on one hand, and rainfall on the other. This would suggest a causal connection between the wheat farming and the soils. For anyone not familiar with this idea, reference to a good atlas which shows the rainfall, vegetation, farming and population patterns of Australia should make this clear at once.

Some geographers have greatly narrowed their field, and in extreme cases have brought geography into disrepute, by limiting their study to causal connections in one direction only—to the effect of the physical environment on man's activities. This shift to determinism will be discussed fully later, but it is now becoming clear that geographers, whether deterministic or not, have too often assumed that the same pattern of distribution of two phenomena proved a causal connection between them. Aside from determinism, this could be in either direction, the distribution of the best alluvium having a causal connection with market gardening, or the patterns of deforestation and overgrazing showing the same pattern as accelerated soil erosion. But even where a causal connection exists, the fact that two maps show the same pattern does not prove this; and it has been realised by some geographers for a long time that the similarities in some patterns are just coincidence.[1]

This assumption of causality may result from incomplete information. When teaching, the author was mystified in two consecutive years to find that thirteen-year-old girls believed in a causal connection between hot climates and oil wells. This turned out to be the fault of the syllabus up to the third form which had included Venezuelan, Persian Gulf and East Indian oilfields, but had not included any oilfields in cold climates. The information available to those girls justified them thinking of a coincidence between hot climates and oil production, but certainly not that one caused the other. In more advanced work there are examples where the lack of information, which would show one phenomenon existing completely independently of the other, or the decision to ignore cases which did not fit the generalisation, have led geographers to suggest causal connections which are not valid.

Jean Brunhes[2] distinguished carefully between causality and connection, giving the following example. In areas of low rainfall we find the smaller, scantier natural or wild vegetation, e.g. the steppes. In particular, the lack of rainfall causes the deserts of the world, the pattern of areas with less than a yearly average of ten inches of rain is the same as the pattern of the deserts, and one defines the other. Now irrigation schemes are to be found in many deserts. Not in all deserts, and certainly not all over any one desert. In his careful wording Brunhes would say that there is a connection between irrigation and deserts, for neither lack of rain nor the fact of the desert itself *causes* the irrigation to come into being. This connection may become less obvious as irrigation spreads into more and more humid areas to increase the yields, but the idea of connection can be modified to state that irrigation is connected with those areas which have insufficient water for optimum crop growth. Different ideas are conveyed by similar words here. Whittlesey wrote of causal connections, while Brunhes used the word causality for this, and the word connection when two phenomena are always found together but one does not cause the other.

A brief glance at two simple ideas of history may make this idea of connection clearer. Two events can happen at the same time, without any connection; Oregon became a Territory in 1848, and 1848 was the Year of Revolutions in Europe, but no historian would see any connection in the mere fact of these events happening at the same time. However, budding historians are warned of the danger of reasoning 'post hoc, propter hoc', and not to leap to the conclusion that because event B happened after event A, it was necessarily a result of A. Similarly, the mere coincidence that two phenomena exist in the same place is not sufficient justification for any geographer to assume that there is any causality or connection between them. It seems that we need at least three concepts to cover the possibilities of several phenomena in the same place: causality, connection and mere coincidence.

Causality between physical topics, in broad outline at least, is clear and accepted. Different rocks cause different types of relief and soils in certain areas. Mountains cause differences in the climate, and the climate has a direct effect on vegetation. Causality in physical phenomena may involve several topics, as tightly folded rocks resulting in the high Andes which modify the climate extensively, which in turn gives different vegetation and soils not only at different heights, but at either side. Even here, more than one factor is involved in each case (e.g., even when a rock 'causes' a scarp, the process of erosion and the length of time are factors too), and when we extend to human phenomena causality is not so direct, one-sided or inevitable. If a social group decides to settle in a limestone area, then the lack of water may cause them to nucleate their dwellings round walls and springs. Much more indirectly, the features of the physical environment are causal factors in how man gets his food. Everyone needs food, and whether man decides to trap, hunt, gather, fish, grow or herd, the rocks, relief, climate, soils, wild vegetation and seas will have some effect, however indirect and partial this may be.

Causality is not completely one-sided. With increasing significance man is changing the relief and modifying the climate. Coastal works, land reclamation, tunnels, quarries, clothes, houses and rainmaking are all tiny starts in this direction. Less clearly understood to the layman are the drastic changes man has made to the soils and vegetation all over this planet. Studies in historical geography[3] show how the green fields we take for granted had to be cleared of trees, drained by the provision of ditches and field drains, be deep-ploughed, limed, marled and fertilised over the centuries. Trees and grass have been removed from millions of square miles, and the area of soil destroyed for ever is fast catching up with this.[4]

Geographical connection is perhaps less familiar because, while there are many examples, quite often the descriptions imply causality rather than state connection clearly. In fact in many cases in substantive works the fact of connection is so obvious and at the same time of so little importance that it is thought not worth mention.

Thus there are innumerable times when two or more of the phenomena, rocks, relief, climate, soil and vegetation, exist in the same place at the same time, connected with each other, but not causally so. There will be rock of some kind under the hill, but the hill is not necessarily a structural feature; the soils may be azonal or intrazonal and have no connection with the present climate and vegetation as have mature zonal soils. In any area, there must be rock, relief and climate, even if the area is devoid of soil and vegetation, and these may simply be connected in the way that the three elements of climate—temperature, rain and wind—are connected in one place without one necessarily causing the other.

In human phenomena, for example, land use and settlement must go together, but in no sense does one cause the other. People settled in an area must either produce food, or do other work in order to buy food, so land use, even urban land use, must be connected with the settlement. Conversely, for the land to be used in any way, there must be people to do the work, and whether they live on the job, commute in five-lane traffic, or trudge ten miles at dawn and dusk, this means settlement in the area. Land use and settlement are connected, but there is no causality. The fact of settlement alone does not cause people to work; the fact of working does not bring the settlement into being. In finer detail, nucleated settlement does not cause people to grow navy beans, nor dispersed settlement to mine diamonds. There are some works which argue that in the other direction pastoral farming causes dispersed settlement and arable farming causes nucleated settlement but there are more exceptions than the rule, and so many other factors involved, yet only two basic types of settlement.[5] More realistically one can say that land reclamation and drainage, whether in ancient Rome, eighteenth-century England, or modern Holland, is connected with a rectangular, gridiron, dispersed pattern of settlement.

An error which creeps into British geography somewhere at the secondary-school level illustrates further this idea of connection. In 'map reading', communications are separated from settlement (i.e. roads from villages) and the kinds of answers which satisfy the examiners, 'this road grew up to join these towns' and 'these villages developed on the road at the foot of the scarp', imply that either settlement or communications can exist without the other. A few glances at the contents pages of advanced works will show that this separation is perpetuated in some of our best works but when the pupil asks 'which came first' one tries to give the picture of the necessarily interconnected development. Geography may separate topics for analysis, but must put them together again. In particular settlement and communications are phenomena which cannot exist completely independently yet don't really 'cause' each other, and form perhaps the most familiar example of geographic connection not often discussed under that title.

In this strict sense there is connection, rather than direct causality, between the location of some minerals and the location of manufacturing. Not all mineral deposits are exploited, but in those cases where they are we find such phrases in the simple school textbooks as 'coal and iron gave rise to a steel industry here'. 'Gave rise to . . .', 'caused the development of . . .', are familiar phrases at this level and we persist in the habit of using them even when the logic is wrong. Men like Abraham Darby sweat blood to force iron ore and coke to give up some iron; the minerals did not jump out of the ground one dark night and cause a steelworks. Again, mining and manufacturing are obviously connected, but not causally so.

In Chapter 15 of *Latin America* Preston James[6] describes how the population clusters in Paraná and São Paulo are high up, well inland, while further south in Santa Catarina and Rio Grande do Sul they are in the coastal lowlands. But then he takes the trouble to give a very important reminder. 'Not until an historical study of settlement shows an actual causal connection between the facts concerning the land and the distribution of people can such a connection be asserted. [To describe correlations] is not to establish any proof that this relationship was actually a motivating force in the minds of the settlers. For involved in this question are . . . all the countless accidents which play such an important part in the irrational course of human events.' This key paragraph by an eminent geographer emphasises three points in this discussion. First, the commonplace occurrence of geographic connection, and the much rarer occurrence of proved causality. Secondly, that James, like so many others who strive to view man's activities objectively, has come to the conclusion that we cannot assume that man always behaves reasonably in the geographic context; man at times is irrational and we must be prepared for that fact; there is no value in giving a reasonable explanation for something which was done without any clear reasoning at the time. Thirdly, and just as important, one may find the fact that two phenomena cover exactly the same area is pure coincidence and that one cannot find any deeper connection than mere coincidence or accident.

Coincidences are more likely to seem important and significant, even when much further research fails to show a connection, in studies limited to a few topics and/or in a small area. The larger the area, the less likely is continued coincidence of two phenomena, while in a total-topic study any one of the sixteen or so topics automatically assumes much less importance. Some coincidences now are seen to have had causal connections in the past. Factories along canals and rivers, now served by lorries, were, of course, at one time served by barges and the location was not coincidental. A necessary practice in geography is to distinguish between the original factors which accounted for the location of an industry or a settlement, and those different factors which have helped firm or town to keep going and to grow. Study of settlements also familiarises one with the idea that functions change while the form or appearance of the phenomena on the earth's surface remain essentially the same. Eighteenth-century, Georgian-style, textile mills in England now produce paper, plastics and electrical goods; the street plans laid out by the Romans are now lined with Woolworths, Halfords, the Co-op and supermarkets.

As James suggests, when the original causality is completely obscured, or so far back in time that the connections may never be satisfactorily established, then we are left now with coincidences on the earth's surface. James would describe the coincident phenomena, relief, climate, vegetation and settlement, in Brazil, but would be careful not to suggest any connections he could not prove. In contrast Derwent Whittlesey[7] would not only avoid unproved connections, but would not even describe those phenomena which did not have any causal connection with other phenomena in the region. In explaining the compage method of regional description Whittlesey was at pains to point out that much of what we find in a region is there by coincidence; the fact that he then ignored these coincidental phenomena need not bother us here.

On this wider scale of full, world-wide, regional description, however, the more obvious coincidences are quite familiar. The major industrial areas of the USA, Britain, Europe, USSR and Japan coincide with the so-called temperate belt of climate. Sydney, in Australia, is in the middle of the main coalfield, but that is coincidence because the coal had not been discovered when

Sydney was founded. The fact that the more favourable half of North America faced Europe, whence came the mass migration of the nineteenth century, was again coincidence. From the evidence presented by Brown[8] of the persistence of settlers in the most difficult places, one gathers that the wave would still have swept over the continent even if the western half had been the more fertile. The coincidence simply made this easier.

While there is at least connection between climate and soils on one hand, and farming on the other, a country with a characteristic climate may also be characterised by a type of mining, or by the fact of heavy industrialisation. The climate and the mining, which may have to be described in the old-fashioned type of regional description, simply coincide in the country. In the full regional description the implication that each topic is closely interrelated with every other topic has been overplayed for too long. This fact, among others, led Whittlesey to adopt another method of description and to call it the compage—'a mere compaction' or collection of topics which happen to exist in one place side by side, at the same time, without any causal connections. Particularly in areas settled for thousands of years we are likely to find these coincidences because of the many changes of function in time. In Britain we have the coincidence of tumuli, earthworks and strip lynchets with the features of modern land use. Even since Roman times, when many of our towns were founded, or modified to their present plan, the changes in farming in the surrounding country-side have been so many and so great that the grouping of one type of settlement with one type of farming is now largely a matter of coincidence.

Sample and example

Some geographers argue that geography, like natural science, should formulate general laws as one of its functions. Other geographers believe that there are so many special cases in geography that it is virtually impossible to state laws which have any value; that in order to cover all the regional variations and exceptions the laws have to be so superficial that they are usually obvious to a child of twelve. Another danger to the formulation of general laws in geography, one which can result in quite erroneous statements, is the habit in research and teaching of making the generalisations first, and giving examples afterwards.

Some readers may be impatient with the many references to geography teaching in this book, arguing that academic research in the universities determines the nature of geography and gives the lead. However, the author is perennially conscious of two things: the way in which the results of new, advanced work are presented to students and the public; and that the present school pupils and college students have among their numbers the geographers of the future.

One sincerely hopes that contemporary geographers examine many specific cases of natural and human phenomena before they attempt to make generalisations and draw up laws, but there is little evidence of this. In the past, geographers have clearly set up a theory and have then selected their evidence to prove it.[9] In the present, however well the research was originally done, too often the books and lectures present the conclusions, that is the generalisations, first, and then give a few examples to back them up. In the author's experience this method of presentation has a disastrous, cumulative effect. First, academics in other disciplines, particularly the older men not familiar with good geography at school, discount both the content and method of geography. Second, the pupils and students exposed to these books and lectures not only absorb some of the subject-matter, they subconsciously absorb and imitate the method as such. The effects are for students to use a poor method of presentation as their method of working and thinking, and for those who later teach and lecture to go on putting the big cart before all the little horses.

Sound methods of research and teaching are supposed to be those of working from the particular to the general, from the known to the unknown, and of examining many cases before stating general laws. But to many teachers and writers it seems better and easier to reveal the grand conclusion first; and having done that, why bother to plod through all the detail which led to this conclusion, why not just give a few examples? It seems that some geography teaching had become so full of generalisations, with no familiar, known, specific cases to give it meaning to the children a few years ago, that there was a reaction in Britain, and a swing to teaching geography by means of sample studies.[10]

While this change from vague generalisations about the Pampas to a detailed examination of one Estancia certainly revived geography for many children, there is still the danger at this other extreme of assuming, or, worse still, asserting that such samples are typical of perhaps thousands of other units. Platt[11] is often praised for his choice of samples which are typical of the regions he describes, but this choice was the result of wide, detailed knowledge, and great skill. For the rest of us, teachers, students and geographers alike, remains the need to be aware of the pitfalls of this particular aspect of the nature of geography. History tends to make much less use of sample and example. Natural science makes much more use than geography, but can do so more safely because natural laws which apply to one specimen or sample apply to all the rest.[12]

If one cannot study every case of a given topic or phenomenon, then one must study a few cases, a few samples. These samples must be chosen to be as representative of the total number as possible, and there are mathematical methods, such as random sampling, to do this.[13] In geographic study, especially of human phenomena, it soon becomes clear that there is so much variation from sample to sample that valid, valuable generalisations are very difficult to make. Moreover, much more important to this argument, is the fact that one sample can rarely be quoted as an example typical of the rest. Much more often in geography one has to quote two or three samples as examples of the different extremes and variations. The reader may be familiar with an old teaching method which first described the general features of youth, maturity and old age of a river, and then named one example. Perhaps teachers, having given the generalisations, have been very hard put to it to find local examples to show their pupils. Much more likely, youthful geographers have stared at scores of rivers in real life, actual samples, and have been unable to fit them into the generalisations. One sometimes wonders just which samples were studied by the great geomorphologists, and at once realises why the same examples are always quoted in the textbooks and exam answers.

Samples used in presentation can be misleading in another way. Experienced geographers would agree that no subject-matter is unique to geography. But many people who teach geography are not trained geographers, and they and young students often believe they are working geographically when they are studying certain subject-matter, whatever their approach to that matter. The phrase 'geographic factor' has been bandied about so often by such people that Simons[14] was driven to explain in words of one syllable that there are no such things as geographic factors, only factors (of any kind) which influence the geography of phenomena, of any kind.

Even Honeybone writes of geographic influences, and of geographic units in the sense of farms and ports.[15]

This being the case, teachers and students can be forgiven for thinking that the study of one farm, factory, town or parish is geography, but the sooner they realise the error the better. The study of one topic in one place is not geography. Even in journals of geography, and certainly in students' studies, we see such titles as: 'The geology of Clenchwarton', 'The soils of Twitteringham', 'Eighteenth-century borax mines round Crumbleton', 'The population of Plockton', 'Little Minting: an urban study', 'The retail structure of St Nectan' and 'Farming in the parish of Wincle'. The misunderstanding is that because geography involves rock, soil, mining, population, towns and farming, then any kind of written work about these topics is geography. For a student, the most difficult thing is to choose an area small enough, or a topic limited enough for him to deal with (having his time and ability), yet still wide enough for distributions, correlations and relationships in space to be revealed.

At best, such sample studies as those mentioned above are on the threshold of geography, in that they form raw material for work with a proper geographic approach. The main value of each study is for comparison or contrast with a similar study done elsewhere. This comparison and contrast may be made by the same student when more research is complete, or by another geographer who collects two or more such studies before him. But the sample study in isolation gives the wrong impression of the emphasis being on certain types of phenomena, on processes of economic activity, rather than on the way *a vast number* of cases are distributed on the earth's surface, related to one another in space, and causally connected with other phenomena.

The later processes of induction, of classifying, and making models and generalisations, all depend for their validity on the accuracy of the individual cases studied in the beginning. Honeybone distinguishes between regional samples and type samples. He states that regional samples 'must be particular, ordinary, and typical of [the] region', while type samples such as an HEP station, glacier or port must reveal the general characteristics. Several times he mentions the necessity of choosing 'typical' rather than 'unusual' examples, but gives no guidance as to how this can be done. A little reflection will show that Honeybone uses the word sample in the sense of example. If one is to choose one regional sample as a typical farm in that region, one must have wide knowledge of many farms to know what *is* typical. Similarly, in using one port to illustrate the characteristics of ports, one must be familiar with many ports.

Two slightly different points therefore need emphasis. First, the many farms and many ports form the samples for research, so that some worthwhile generalisations may be made. The word sample, obtained by methods of random or stratified sampling, is used in that sense here. Second, one farm or one port may be described in one's final presentation as an example of the many individuals studied, but one example must not be used for research, and it is important to make clear in the final presentation just how many samples were studied, and how they were chosen. One of the most valid criticisms of past geography is that samples were chosen in a most subjective or biased way, often to prove rather than test an hypothesis.

Models

One of the most confusing changes in geography is the changing use of words. In some cases new words are used for old ideas; in other cases old words are made to carry more and more meanings simultaneously. Perhaps it is too much to expect the new words to be used only for new ideas and techniques. Burrill[16] is most concerned about overloading words, and using the wrong words so that geographers can neither classify accurately nor communicate coherently with other disciplines. In particular he criticises those advanced workers who try to increase their prestige by inventing new jargon. 'The failure of communication between the users of these languages and the rest of the geographic profession is profound, and the inability of the uninitiated to grasp the concepts and ideas of the initiates is serious.' James,[17] Johnston[18] and Morgan and Moss[19] refer to similar problems in slightly different contexts, and, of course, one function of general systems theory is to try to reduce these problems.

Because of this play with words, many people assume that the new word model is just the old word generalisation. Moreover, in some recent works which use the word model, model-making seems to be the end purpose of geography. Haggett[20] certainly gave this impression whether he intended to or not. In other works, models are seen as yet another means to the end of comprehending what goes on on the earth's surface.[21] A conclusive study of model theory must be deferred because there is not space here, the theories are still developing and changing rapidly, and because some of the advocates of models seem to have misunderstood their colleagues to the extent that there are contradictory explanations of what the word model means.

Cole and King[22] list many meanings of the word model, and a modified list is given here. Thus different authorities assert that a model is an analogy, an ideal, a representation, a description, an abstraction, a generalisation, a frame of reference, a programme of research, an explanation of how a system works, a theory, less than a theory, a demonstration. But common to their definitions is the basic idea that a model is a simplification of reality. Cole and King go on to explain the types of models:

(1) Scale models, including maps.
(2) Simulation models—mathematical probability.
(3) Mathematical models—mathematical certainty.
(4) Theoretical models which give concepts or frameworks to
    (a) formulate problems
    (b) suggest solutions.

To paraphrase and supplement Haggett,[23] we make simplified representations of reality in order to try to understand it and then demonstrate those properties which we have understood. Haggett, too, stresses that a model of any kind is a simplification, and one which provides a working hypothesis to test against reality. He lists six types of models:

(1) *Iconic,* three-dimensional models smaller (a globe) or larger (a model of a molecule) than the real thing.

(2) *Analog,* where one phenomenon is represented by another, more familiar.

(3) *Symbolic,* representation by mathematical symbols.

Iconic models would seem to include working or experimental models; analog models would seem to include what Haggett calls natural models where the growth of a town is likened to the spread of an ice cap. Similarly symbolic models would seem to be mathematical models.

Geographers may borrow models from other disciplines, such as the gravity model or the refraction model. Others, such as von Thünen, and Christaller, started with the very barest over-simplifications and gradually increased the complexity. In contrast, others, such as Taaffe in his study of Ghana's transport, start with the complex reality and simplify more and more. 'The problem in each case is to translate the circumstances being studied into some analogous form in which it is either simpler, or more accessible, or more easily controlled and measured.'[24]

If the word model is vague, then it cannot be surprising that when geographers have borrowed or built their own models, and then worked on them, the purposes and applications vary enormously. A good model can be intended as an ideal for planners to work toward; a finished description of reality; a generalisation, based on a model of one part of the world, which will apply to the rest of the world; a standard against which to compare regional variations; an explanation of reality; an hypothesis to be tested. Haggett cheerfully admits the contradictions, the crudity and the over-simplification of many models at the moment. At least he is aware of all the shortcomings. While the basic value of models is apparent, one must take notice of a few serious faults which the enthusiasts have been able to ignore so far.

There are many implications that good models are universals. If this were so, then Christaller's models should apply to China, India, Brazil and so on, as well as to southern Germany, East Anglia and the Midwest. More seriously, one doubts whether the variates and attributes used to arrange Western towns in a functional hierarchy can apply in other cultures. The more a model attempts to be a generalisation, the simpler it must become. There are dangers in both extremes: of one model which will apply to every case of a world-wide topic, so erasing regional differences; and of so many complicated models that there are as many models as real cases. Haggett[25] suggests an approach to regional geography might be to examine regions as special cases of a general model. Now, does one make the model first, then study regions, or study the regions and then make a model, based on evidence?

This seeming universality of models is not as well discussed as the necessary simplicity. The essential point of any type of model is that it is simpler than the real thing. For these simple, general models to be applied to real cases, the mathematical ones in particular have to be weighted in many different ways. The gravity model which is used to study traffic flow between towns has to be weighted to allow for such factors as type of transport, cost of travel, individual preference, wealth of the populations, economic levels, technological levels, habit and custom, time of year and so on. Here we see the clash between the new quantitative and old qualitative geographers. Some of the older geographers must be genuinely astounded that some young researcher can calmly hope that a simple formula like $m = \dfrac{PiPj}{dij}$ where Pi and Pj are the populations of towns i and j, and d is the distance between them, can describe and explain all movement (m) between towns.

The tragedy here is that just as some of the older geographers cannot accept all these new techniques at once, too many of the newer people discard the old qualitative techniques completely. The word techniques is used intentionally; geographers may have lacked mathematical precision, but they were trained to take account of *all* the factors in a given situation. One realises that a model which becomes too complicated loses its value, but instead of a revolution—qualitative work out, quantitative in—one could hope for a combination of the two techniques, so that the new mathematicians will not have to re-learn the attitude of mind which made the earlier artists aware of all the myriad factors, measurable and intangible alike. Earlier generations may not have been able to measure and compute the things they described, but mathematical techniques will be no gain at all, if most of the geographer's other skills are discarded.

1. Whittlesey, D., in *American Geography: Inventory and Prospect*, Eds. James, P. E., and Jones, C. F., Syracuse University Press, New York, 1954, p. 25
2. Brunhes, J., *Human Geography*, Harrap, 1956, p. 23
3. Brown, R. H., *Historical Geography of the United States*, Harcourt, Brace & World, 1948
4. Jacks, G. V., and Whyte, R. O., *The Rape of the Earth*, Faber, 1949
5. Houston, J. M., *A Social Geography of Europe*, Duckworth, 1953, ch. 5
6. James, P. E., *Latin America*, Cassell, 1959, pp. 509–10
7. Whittlesey, op. cit.
8. Brown, op. cit.
9. Semple, E. C., *Influences of Geographic Environment*, Henry Holt, New York; Constable, London, 1911
10. 'Sample Studies', *The Geographical Association*, 1962; Rushby, Bell and Dybeck, *The Study Geographies*, Longmans
11. Platt, R. S., *Latin America*, McGraw Hill, 1943
12. Toulmin, S., *The Philosophy of Science*, HUL, 1953

13. Gregory, S., *Statistical Methods and the Geographer*, Longmans, 1968
14. Simons, M., 'What is a Geographic Factor?', *Geography*, vol. 51, no. 3, November 1966, p. 210
15. Honeybone, R. C., in 'Sample Studies', *The Geographical Association*
16. Burrill, M. F., 'The Language of Geography', *AAAG*, vol. 58, no. 1, 1968, p. 1
17. James, P. E., 'On the Origin and Persistence of Error in Geography', *AAAG*, vol. 57, 1967
18. Johnston, R. J., 'The Subjectivity of Objective Methods', *AAAG*, vol. 58, 1968, p. 575
19. Morgan and Moss, 'Geography and Ecology', *AAAG*, vol. 53, 1963, p. 1
20. Haggett, P., *Locational Analysis in Human Geography*, Arnold, 1965
21. Walford, R., 'Operational Games and Geography Teaching', *Geography*, vol. 54, 1969, p. 34
22. Cole and King, op. cit.
23. Haggett, op. cit., p. 23
24. ibid., p. 21
25. Chorley, R. J., and Haggett, P., *Frontiers in Geographical Teaching*, Methuen, 1965, p. 364
For models see also Chorley, R. J., and Haggett, P., *Models in Geography*, Methuen, 1967; Chorley, R. J., 'Geography and Analogue Theory', *AAAG*, vol. 54, 1964; Harvey, D., *Explanation in Geography*, Arnold, 1969

8

FUNCTIONS AND PURPOSES OF GEOGRAPHY

The student familiar with the content and techniques of geography must think at some early stage about the geographical approach and how it is distinct, say, from that of history. It would be fascinating to know just how many students and geographers are concerned about the purposes of all this work, beyond academic interest, and at just what stage this concern becomes obsessive.

Interest in the facts of the earth's surface, and in the techniques used to study it, must be put first. Yet even this interest will vary according to the person. The geographer is interested to carry out new research; the student to know what has been found out. Moreover the interest of a student intending to become a geographer will be different from that of students with other vocations, reading geography as part of their course. While some overall philosophical purpose of a discipline may not be the main attraction to a young person, it seems vital to consider this question for several reasons:

(1) To distingiish the purposes, functions and short-term uses of geography.

(2) To examine whether individual aims of geographers contribute to some general aim.

(3) To make clear that many geographers have some very odd purposes.

(4) To help to decide whether geography has any constant purpose.

(5) To help make clearer the nature of geography to all those interested in it, and

(6) To show the novice that the purposes of research can be completely different from the purposes of geography in schools.

This chapter and the next are not intended to be conclusive, rather to give an introduction to these points as a basis for further study and thought.

It has been suggested earlier that purposes and functions are often very difficult to distinguish. One worker sees it as the end purpose of his work to construct a model or generalisation, or simply to describe. Another worker uses models simply to help in analysis, and attempts to explain as well as describe. Straightforward geographic description has at times been the purpose of geography, then a function in the process of wider analysis, and at the moment is considered one of the least important aspects of the geographer's work. While men such as Varenius, Humboldt and Ritter attempted to analyse and reduce to order the complexity of the earth's surface, their first lengthy step was to attempt a complete description. Humboldt's aim was complete knowledge of the earth's surface, which might lead him to a philosophy. Ritter considered similar factual work to be a necessary preliminary to his study of history. In more recent times, description of smaller areas has been the purpose of the French geographers who wrote the regional monographs of the *pays* of France.[1] Even in the advanced texts produced in Britain in the first half of this century description, often mainly of physical features, has formed the bulk of the work, if not its ultimate purpose.

Geographical description is going out of fashion for several reasons. As a necessary part of regional geography it has suffered in the decline of that branch of the discipline which usually offered such superficial explanations and analysis as to be intellectually unsatisfying. Moreover, there is little interest in the study of the correlations and causal connections of all phenomena in the landscape at the moment. Description, a very difficult art as it is, could never be a completely satisfying purpose in itself, but it would be a serious loss if it disappeared completely. Many geographers today claim their purpose is to study distributions and spatial relationships, but if they as professionals confined themselves completely to this, the loss to educated laymen would be serious. For, whether it is the purpose of geography or not, one of its most persistent and worthwhile functions has been to present an orderly body of knowledge about the earth's surface, invaluable in education and in our general culture.

We have a curiosity about the past, about how people lived and how events led up to our time. We have a curiosity about things surrounding us, flowers, animals, rocks and machines. We have a curiosity about distant places, partly satisfied by travel and travel books, but better served by good geographical description. One may want to understand why one of those dreadful west Yorkshire valleys is like it is, and how it fits into the wider scheme of things beyond the moor, over the Pennines and past the coalfield. The student of geography wants exact knowledge and a complete world picture, rather than the impressions and titbits of the tourist. For example, one sees sheep by the roadside, but finds from the statistics that cereals cover a larger acreage and are more profitable. Or, in a farming area, one finds that people make more money from tourism. Hollywood suggests that barn dances are a feature of the western USA; the facts of rural settlement show a most depressing lack of social life and amenities. The student of geography reveals himself in the field as the person who wants to get an accurate picture of the world around him, and a full picture of the landscape beyond the hill and the horizon. For one man on foot this is impossible, and hence the continuing necessity for the armchair and the geographic description, produced by disciplined investigation and thoughtful presentation.

Both Houston[2] and Freeman[3] write of the encyclopaedic element in geography, Houston as a stage which has passed, Freeman as one of six continuing trends. Hartshorne has insisted that if geography were no more than a catalogue of facts about the earth's surface, it could not rank as an intellectual discipline, yet, at some stage, the facts which geography presents are an essential part of our knowledge of the universe outside our bodies. No other subject or discipline provides the type of information on which geographers carry out their analytical operations. Thus our factual knowledge of the complete landscape as distinct from compartmentalised knowledge of rocks, climate, flora and fauna, buildings and field boundaries comes from books of geography. Mention will be made later of the idea that the purpose of geography is to study not any particular separate phenomena, but to study the phenomena as they are found in the landscape, and to analyse their combinations and relationships as they exist on the earth's surface. Geography, unlike geomorphology, is the discipline which, for example, describes the shape of a particular area of the earth's surface, and shows how landforms are distributed and arranged one in relation to the other.

It was suggested earlier that geographers are forced to study certain economic and social activities in the absence of other disciplines. It is certainly a function of geography, which no other subject performs, to pass on this knowledge of economic activities. The idealist would insist that the study of how men catch herring and cod, how they farm, how textiles, cars and chemicals are made, how iron ore and coal are mined, is not geography. The pragmatist would state that geography is the only subject in the normal British school syllabus in which this type of factual information is relevant; and there is no doubting its fascination to children and students alike.

In the older books which never mention central place theory, gravity models and the like, the main information about towns is of site, situation, form and function. One admits that towns have been studied for too long as unique cases, and that explanations of the form and function have resulted in lengthy historical accounts starting with the Romans, which certainly are not geography. But one would prefer to see the accounts of street plan and historical development combined with the more geographically pure studies of functional zones and urban fields, rather than discarded and ignored.

Another overworked word these days is problem. It is normal for problems of geography to be combined and confused with the daily problems of people in some particular part of the world. Thus a problem of the population geographer might be 'Why are these people located here and distributed in this way?' The problems of the population itself might be 'How are we to make a living, or prevent overpopulation?' Yet it is commonplace in geography to find the geographer concerned not only with the problems of analysing and explaining the distribution, density and location of population groups, but also identifying himself with them in his concern about water supply, soil erosion, deforestation, mineral exhaustion, crop failure, provision of infrastructure, development of HEP, family planning and international agreements. This is at one and the same time a criticism of the wooliness of geography and a statement of the fact that geography is the medium through which the educated layman is made aware of such things as resource development, flood control, soil conservation, international trade, underdeveloped countries, power and irrigation schemes, industrial decay, the population explosion, disease and poverty throughout the world. These are not problems of how phenomena are distributed, located and interconnected on the earth's surface, which are the intellectual problems of the geographer. They are vital problems of the people living on the earth this day, but problems brought to our attention by the journalist, politician and geographer, rather than by the historian or natural scientist. So the nice distinction must be made between bringing these problems to the attention of educated people, one function of the geographer, and solving the problems of irrigating a farm or building a harbour, which is the function of the civil engineer.

Briefly to mention the use of geography, one should simply point to its value in education and in adding to human knowledge, without being precious or labouring the point. But there are those nowadays who insist on a more direct, tangible and profitable outcome from an expensive education. There are very few careers for geographers outside teaching and academic research. The use of their training to geography graduates in other careers can be measured only by their happiness and success. Geography is clearly of less material use than natural science either in getting a job where one can make direct use of one's geographic knowledge and approach, or in materially and financially helping mankind. But geography and geographers are becoming increasingly useful to the infinite variety of planners who hope to organise better landscapes, economy and societies for us in the future.

First, the practical techniques are not peculiar to geography; even more, the geographer is not as skilled in making maps as is the surveyor or cartographer. There are no practical, technical, manual, geographic skills which the geographer can go out and 'do' for society. These are better done by the geologist, surveyor, meteorologist, pedologist, botanist, agronomist, mining engineer, civil engineer, architect, cartographer, etc., etc. However, the geographer has a point of view, a method of study, a special type of knowledge, call it what you will, which makes him best qualified to advise on certain aspects of our material life. Because the regional geographer studies and understands the interrelationships between diverse phenomena in an area, because the general geographer studies and understands regional variation from one part of the world to another and sees the total extent of one or more phenomena, they are in the best position to advise on planning and development in any economic activity.

In this light, it seems that the application of geography is becoming more important because even well-developed countries now try to plan and develop rationally. Moreover we begin to admit that changes in one phenomenon affect many others (e.g. improved colonial hygiene leads to increased population which leads to starvation and unemployment). The under-developed countries must develop rapidly, cannot afford serious mistakes, and the geographer can suggest a plan, especially as in over-crowded countries farming, forestry, mining, industry, transport, housing and recreational facilities must all be co-ordinated. Geography, as an academic study, has long been concerned with just such topics. The geographer is not trained to go out to plant trees, reorganise farming, find mineral desposits, site dams, plan towns, start factories or build roads. But he is equipped to do several vital things: first, to supply factual knowledge as the foundation of a plan, or to find this out in countries with little published geography; second, to draw attention to, and keep attention on, the necessity for balance in food supply, manu-factured goods, social services, water supply, roads, open spaces, recreational facilities and so on. Too often agricultural schemes have ignored the relief, climate, soil[4] or wild life.[5] Town planners forget to provide pubs, playgrounds, bingo halls or even garages;[6] or a city is enlarged by several thousand houses without a corres-ponding increase in water supply, sewage farms, rubbish dumps and roads. The geographer is in the habit of thinking of the actual interrelation of such diverse things, and of realising at once that a change in one will start a chain reaction demanding or resulting in changes in many others.[7,8]

Thirdly, certain individual geographers may be able to draw up the overall plan for economic and social development, better land use or whatever. But there is still much argument as to whether the majority of geographers are able to do this, even if they want to. Many laymen and geographers insist that a team of experts with a wide variety of skills is needed. Moreover, there are some who argue that geographers should not take part in practical schemes but should remain purely academic. For the moment, then, one may say that the geographer is neither a technician nor an engineer of any kind. But he is practical in the sense that he studies the world as it is, complex, complicated, often irrational and unplanned, badly organised and certainly often inconsistent. As much as any other scholar he has to face up to reality where it matters to the ordinary man and woman here, now, trying to make a living and bring up children. While the historian studies what once was reality, and the chemist studies what one day may have practical application, each studies something isolated and remote from ordinary everyday life. The best geographers do not shun the immediate, the human, inconsistent and messy, they strive to see it as it actually is, and to understand it. From this start, they are then well prepared to help planning with two vital contributions. They can draw attention to all factors and all interconnected phenomena however unexpected and seemingly unconnected they are to the layman. They can also help to shape plans which will work in real life, and not just on the architect's drawing board. To this end they are constantly aware that people often want television sets before washing machines, bicycles before schools, fish and chip shops and pawnbrokers before civic theatres.[9]

Two last points on the use of geography in planning. Many statistical techniques used in other disciplines and now adopted by geographers help one to predict what will probably happen in the future. These probability forecasts are completely separate from planning the economy, cities and social amenities which we would like in the future. Second, this planning is not geography; it stands in relation to geography as technology stands in relation to pure science. Experts in geography can be very useful in planning, but they do geography a disservice when they insist that the whole purpose of geography should be to plan our future economy and landscape. This is as short sighted and mistaken as insisting that all research in natural science be directed to solving immediate material problems.

It is so easy to be mistaken about the nature of geography from an examination of its content, method and techniques that Lukermann has stated that its nature must be defined by the questions it asks.[10] Thus we may have to state what geographers set out to understand, or what they ought to try to understand, to define the discipline. Some people seem to be working in geography when they study landforms, farm economy, industrial history, city architecture, because these phenomena are also the subject-matter of geography. But such studies are more properly geomorphology, economics, history and so on, obviously enough.

Others seem to be doing geography when they study a country, or part of the earth's surface; but many so-called regional geographies are little more than gazetteers on an areal basis and make little attempt to analyse distributions, make careful areal correlations, or to explain the spatial relationships. Any educated person prepared to dig out the facts could do the same without any special geographic training. Similarly the results of prolonged fieldwork, perusal of the maps, or advanced mathematical calculation are not necessarily geographic, that is do not reveal the geographic approach outlined in Chapter 4. What is needed in addition to subject-matter, approach and technique is a clear understanding of why one is going to all this trouble.

It is difficult to agree with Lukermann that the basic question is 'Why do we see the world divided into regions?', yet Hartshorne's definition of geography as the study of 'the areal differentiation of interrelated phenomena on the surface of the earth of significance to man' implied this same contrast of one part of the world with another.[11] However, Lukermann is more concerned with the kind of understanding which genuine geographical study gives us. To paraphrase him briefly, it gives us understanding of how all phenomena fit into the complex system on the earth's surface. We gain insight into:

(1) Location, that place affects our actions; 'if the geographer fails to conceive of his research as involving this step it seems difficult to accept such study as either explanatory or descriptive in any geographic sense'.

(2) Distribution and localisation, i.e. concentration of phenomena in certain places.

(3) World-wide interdependence of physical and human systems.

(4) Interaction in place between phenomena isolated for study by other disciplines.

(5) Dynamic change in places, phenomena, and the importance of resources.

(6) The cultural landscape, i.e. in the American sense of the landscape as we see it today, a combination of physical features and human structures, the present stage in the ceaseless interaction between man and nature.

Finally, Lukermann states 'The geographer's purpose is to understand man's experience in space.' This sums up what the present author believes, but it would take more than one book to deduce all the ways in which the geographer might go about this, and it might surprise the sixth-former studying geography to learn that this is the purpose of the research worker.

By examining both what geographers have set out to understand, and the implications of earlier chapters in this book, one may list the kind of studies proper to geography. There is some duplication resulting mainly from the different practical approaches of the regional and general geographer. Geography as a whole aims to understand:

(1) The distribution of phenomena on the earth's surface.

(2) Their connection with other phenomena in the same place.

(3) Their relationship to other phenomena in adjacent or distant places.

(4) The effects of one or more phenomena on others.

(5) Variations of a phenomenon from place to place.

(6) Why phenomena exist in certain places and not in others.

(7) The spatial diffusion of phenomena.

(8) Reciprocal movements of phenomena.

(9) The spacing of discontinuous phenomena.

(10) The shape of networks.

(11) Density and grouping of discontinuous phenomena.

(12) Location and localisation of phenomena.

(13) The confinement of certain peoples and activities to certain places.

(14) The effect of actions in one place on actions in another.

Up to this point geography could be dry, abstract or merely trivial. Studies of the above topics could result in us having only such knowledge as that peaks are usually surrounded by corries, people live in fertile areas, the Danes possess Jutland peninsula, or sewage works are located outside the town. If this were all, geography could degenerate into a geometric formulation of the patterns on the earth's surface. The basic questions common to the topics above are: How is each phenomenon arranged on the earth's surface?; How are phenomena interconnected on the earth's surface?; What is the sum total, and what are the inter-relationships, of phenomena in a given area (region)?

By asking different questions, but still within the geographer's terms of reference, we can get nearer to matters of vital concern to man.

D

(A) *The earth's surface*

What is it made of?

What shape is it?

What are the climate, soils and vegetation like?

What do the people do for a living?

How are the people grouped?

Where and how do the people live?

What are they like, what are the characteristics of the population?

Which parts of the earth belong to which people?

And, in each case, why?

(B) *Regions*

It is self-evident that one part of the earth's surface is different from another in the features listed under (A).

How do these features differ?

Why do they differ?

Are the differences significant to man or of merely intellectual interest?

Are there recurring patterns, so that we can classify and generalise?

Are the systems basically similar so that in understanding one we understand them all?

The fundamental questions common to (A) and (B) are: What are other parts of the world like and how do we, here, fit into the complete system?

(C) *The world*

The descriptions and explanations which form the answers to these questions provide a body of knowledge about:

The world we live on.

How we use it and live on it, and this in turn poses further questions.

Is this the only way to live?

Are we forced to use the earth this way?

Is there a better way?

So we are led to question human experience in space, as Luker-mann says. We are led on to try to understand our relationship with the earth, and earth's advantages and drawbacks for human life. Also, how we depend on a certain amount of surface-space in order to live, and yet at the same time have to overcome problems of space and distance. Ultimately, we may be able to arrange ourselves on, and use, the earth's surface in a better way.

1. From about 1900 onwards. For details see: Taylor, G., Ed., *Geography in the Twentieth Century*, Methuen, 1957, p. 76
2. Houston, J. M., *A Social Geography of Europe*, Duckworth, 1953
3. Freeman, T. W. *A Hundred Years of Geography*, Duckworth, 1961
4. The East African groundnut scheme
5. Rice at Humpty Doo, Australia, destroyed by the birds
6. Brown, R., 'City folk find new town so friendly', *The Guardian*, 10 September 1968, Cumbernauld
7. Freeman, T. W., *Geography and Planning*, HUL, 1958
8. Stamp, L. D., *Applied Geography*, Pelican, 1960
9. Bendixson, T., 'Professor attacks naïve urban planning criteria', *The Guardian*, 9 September 1968
10. Lukermann, F., 'Geography as a formal intellectual discipline, and the way in which it contributes to human knowledge', *The Canadian Geographer*, vol. 8, no. 4, 1964, p. 167
11. Hartshorne, R., 'The Nature of Geography', *AAAG*, Lancaster, Pennsylvania, 1939, ch 11

9

DETERMINISM

Just as each individual geographer can study only part of the earth or a few phenomena at any one time, so the purpose of a particular piece of work must be less than the purpose of geography as a whole, but can contribute to it. However, there have been individuals and schools of geographers who have persisted in attributing some very limited and bizarre purposes to geography. Some will be mentioned briefly, but greater attention will be paid to the determinist school because of its importance to our main theme.

In the second half of the nineteenth century it was believed not only that geographers should confine themselves entirely to physical phenomena, but also that it was possible to observe a 'natural' landscape completely untouched by man. Nowadays we are aware, at least, of man's extensive interference in the soils and vegetation of the world. Other schools occupied themselves with the observable or cultural landscape, taking account of man's activities only where they resulted in physical changes in or additions to the natural landscape. This certainly was a partial study of the earth, but at least it put the emphasis on landscape and the appearance of the real world. The emphasis in the late nineteenth and early twentieth century was on the *content* of geography, the material landscape. Increasingly we see an emphasis in the second half of this century on the *method* of geography, on the analysis of distributions, arrangements and relationships. Carried to extremes, this becomes an exercise in geometry and statistical analysis which idealises spatial arrangements until they seem to have little connection with the earth's surface. This new type of work is just as partial as the old, and the answer lies in a fusion of these two approaches with others.

Other partial purposes are, perhaps, more familiar. One defence against the criticism that geographers study phenomena studied more 'scientifically' by other disciplines, is that the geographer studies them just as they are on the earth's surface, not isolated from their environments. Some geographers have seen it as their purpose to study the distributions, associations and interrelationships of a hodge-podge of topics just because they all happen to occupy the same area of land. At one time it was believed that all phenomena in one place were causally interrelated, but now we are beginning to realise that this is not so. 'Geography, because of its very nature, cannot compete with other subjects purely on the basis of systematic specialisation . . . geography either stands or falls as an integrating discipline.'[1]

As recently as 1964 Lukermann could state that the purpose of geography is to study the regions of the world.[2] This as an end in itself now seems as old-fashioned as the theme of school geography between the wars, the study of 'Life and Work in Many Lands'. There were times when the purpose of geography must have appeared to some people to be compiling lists of longest rivers, highest mountains, capital cities, routes of railways and so on. Certainly it had its propaganda value in emphasising the importance of Britain and the British Empire by means of lists of imports and exports, and great red patches in the atlas. There are still geographies of North America which devote more pages to Canada than to the USA. This paragraph may seem a bit quaint and pointless until we realise that geography is still being exploited for ulterior motives. 'The aim of any selective teaching of geography must be to concentrate on the problems which men need to solve so as to provide for increasing numbers and a higher standard of living. . . . It is the business of well-conceived geography teaching to give children a balanced appreciation of world problems.'[3] These may be laudable aims, but if teachers carried out these orders to the letter, the next generation of undergraduates would have no idea of the nature of academic geography at all. In addition to propaganda for the Empire and World Cooperation, even religion was brought in when Guyot (1807–84) started courses 'showing the harmony between natural sciences and revealed religion'.

Hartshorne believed that the dualism between physical and

human geography was the most serious threat to its continued existence as a discipline.[4] This dualism finds it expression, among other things, in contrasting purposes of geographers. Sauer and Barrows have gone to opposite extremes, Sauer tending to the study of the physical geography of landscapes, eliminating man, Barrows seeing geography as human ecology and tending even to eliminate the physical environment.[5] Geographers have either concentrated on one of the two branches physical and human, or have shown an unnecessary concern with the impact of one on the other. G. P. Marsh (1801–82) was concerned to show how man had destroyed the earth through soil erosion, deforestation, floods, mining and so on. He saw the purpose of geography as the study of what man has done to the earth. Later, Bryan wrote of 'the modern conception of human geography as a study of the adaptation of nature or natural environment by man in the process of satisfying his desires'.[6] In contrast Ratzel, and so many others since 1882 when *Anthropogeographie* was published, set out to prove that man is a product of the earth's surface, determined by physical laws.[7]

The importance of determinism has been argued about in geography for nearly a century now. Because of this, because laymen and teachers still think geography is deterministic, and because two vital recent papers help us to settle the question, determinism will be examined in some detail here. As any philosopher would state that the question of environmental determinism applies not to geography alone, the questions which arise are:

(1) How did geography become so concerned with the problem of determinism?

(2) What has been its effect on geography?

(3) What made geographers think they could solve this problem?

(4) Is the examination of determinism the purpose of geography?

(5) Is determinism, as many geographers believe, now irrelevant to geography?

The question of whether man's actions are determined entirely by his environment, leaving him no free will, is one of the basic philosophical questions. Geography has always been closely connected with philosophy, especially since the time of Kant (1724–1804) who lectured in physical geography[8] in Königsberg. It seems presumptuous however for geographers to have adopted this problem as their own. Certainly their studies, involving man and his physical environment, can provide facts for the philosopher, but three points need emphasising. This can be a function of geography, not its proper purpose. The physical environment is only part of man's environment and even if we could prove that man is not determined by his environment, this does not prove that he has free will—there may be some other determinant.

Possibly Humboldt and Ritter were concerned with determinism, as the question is implicit in Humboldt's *Kosmos* which studied the 'relationships between the physical and intellectual world',[9] and in Ritter's study of geography as preliminary work to his study of history. However, the involvement of geography with the question of environmental determinism was made explicit by Ratzel, followed up by Semple, and has been with us since. Even without the particular works by Ratzel and Semple, it seems that geography must have become involved with the question of determinism for several reasons.

While history studies man and natural science the environment, geography studies both. Ritter, Semple, Buckle and many others have believed that the geography of an area helps to explain man's historic actions there, even if it did not determine them. Some historians of geography say that geographers strove to make historians aware of the 'geographic' factor in history; others that historians seized on the facts of the physical environment and used them so successfully in their explanations that geographers followed suit. Buckle[10] was a pioneer in this field. After the publication of *Origin of Species* by Darwin in 1859, which showed how plants and animals have adapted to their environments, it seemed logical, to some, to assert that man, as an animal, must have adapted in the same way. The reciprocal action of the plants, animals, and to a much greater degree man, changing the environments in which they live, is only just now being understood. Those who have tried to make geography an exact natural science have needed to look for cause and effect, and regular, predictable behaviour in the phenomena they study. Not for them the unpredictable human actions of history; they needed to find laws as precise as those of physics. Thus one tendency in geography has been to generalise about the similarities of human behaviour in different parts of the world, and, at the crudest extreme of scientism, to write only of physical stimulus and human response.

Even the most careful geographers have at times come close to slipping into determinist ways of thinking. As geography studies a great variety of phenomena in position on the earth's surface, and as a function, considers the causal connections between them, there is a tendency to select only those phenomena which are causally connected. Whittlesey was well aware of this danger in selecting topics for the compage. If, from total reality, one selects only those phenomena showing causality and connection, then one is likely to end up with a biassed account which seems to prove that man is adjusted to his environment very nicely. Brunhes admitted his bias in this direction when he made it clear that he would exclude from human geography all features not connected with physical phenomena.

Other geographers have followed Brunhes' lead, and geography teachers, especially those not trained in geography and having to follow suit rather than think things out for themselves, have followed the safe path of including only those human features directly connected with the physical environment. This strict adherence to 'geographic' topics and 'geographic factors' is one of the worst aspects of geography in British schools today.

However, another aspect of geography leads to the same result. If we insist on the use of maps, then there is a tendency to exclude every factor which cannot be mapped and correlated with physical features. So often in Britain and Europe a full explanation involves extensive reference to historical factors in geography. With the less able and more rigid geographers and teachers there is a tendency to ignore these 'non-geographic' factors, and to ignore all social, economic, political and historical factors which cannot be shown on a map. Thus the geography is biassed, one-sided and incomplete.

It will be emphasised later that geography selects its topics and studies them according to their significance to man. This is a bias necessary both to limit the field and to give the unique point of view, but if pursued thoughtlessly again leads to determinism. Thus we study physical features as they affect man, and we view the physical world as the environment of man. We might select different topics, view them a different way, and have a different emphasis if we were to study the world as the home of fish, birds or cockroaches.

Hartshorne[11] gives the impression that he sees determinist thinking as responsible for the crippling dualism of geography as physical and human. However, there is some evidence that the separate growth of geomorphology and the social sciences were responsible for the rift, and this rift in turn has encouraged determinist thinking.

At its best, geography considers the relationships between many phenomena in the total region (see Table 1, p. 52), and between any pair. It studies climate in relation to relief, and population in relation to land use just as much as it considers soils in relation to settlement. It does not necessarily consider one physical topic in relation to one human topic. At its worst, geography considers only the effect of physical features on man's activities and leads to determinist conclusions. Those geographers who still perceive a division into physical and human, and think geography is a comparison of the physical with the human, are most likely to persist with an unconscious bias to determinism.

Therefore it is suggested that the identification of geography with the philosophical problem of determinism was inevitable, sooner or later. The staggering point is that many geographers have believed that they can answer such questions as 'to what extent are man's actions determined by his physical environment?'. Teachers still set out to show children man's adaptation to his environment; for example, the GCE syllabus in a grammar school in 1969 read: 'Geography of Europe—to study the influence of rocks and climate on the people.' One would suggest that if some of the most famous geographers who thought they could answer the question lacked the necessary discipline in logic and metaphysics, then so do most geography teachers.

Demolins, influenced by Le Play and de Tourville, and in turn to influence Geddes, published in 1903 *Comment la route creé le type social*. The word route was used in the widest sense to mean the route over which any group of people had migrated *and* the areas in which they had paused and in which they live now. Demolins' insistent point was that the physical conditions had determined both the type of work and the social organisation of the people. He reasoned that race (and men's ideas) could not be a factor in forming a society because he could find no cause of the different racial types! This is his inability, not lack of a cause. Therefore, he argued, nature (and physical conditions) is the only factor in forming societies. The fact that he could not trace physical phenomena back to first causes did not seem to trouble

him. As a further illustration of his very poor logic, Demolins did
not search for facts to ascertain the route of migration of a particular society; having some facts of the social characteristics
before him he stated where the people *must* have come from,
whether in fact they did or not. For example Demolins said the
Chinese must have come from Tibet. He seemed to set out to show
by induction that environment has an effect on people. In fact he
assumed the effect, then deduced what kind of an environment
might have produced the people he knew.

   This misuse of induction is common in determinist geography.
For inductive reasoning to be used properly, one must observe the
facts, then form a generalisation which will fit all observed facts.
Too often such geographers have *a priori* set up a generalisation,
then set out to prove it by selecting facts which support it and
ignoring all the others. Buckle set out to show how geography
had controlled history, not to see if or not. Demolins set out to
show how routes and places had determined societies. Semple set
out to show that man is determined by nature. Writing of man,
she states, 'the earth has . . . set him tasks . . . directed his
thoughts'.[12] Her work has been criticised as 'glibly enunciated,
superficially exploited, and compelled to support conclusions
without adequate inductive material and in disregard of other
types of causation'.[13] The contrived reasoning must make some
philosophers smile. For example, Buckle insisted that man's
actions are determined by nature. Later he contrasts man's victory
over nature in Europe with his subjugation everywhere else.
Realising this contradiction he explains that nature determined
that man should take control of his environment in Europe, i.e.
it was determined that he should not be determined.

   The confidence of these people at the turn of the century is even
more amazing when one realises that they attempted to answer
the question 'to what extent is man controlled by his physical
environment?' at a time when sufficient data and precise techniques were conspicuously absent. Even today, with radio, news-
papers, periodicals galore, maps, air photographs and a deluge of
statistical data, it still happens that one cannot obtain the precise
data for a certain area in the form which one requires. Only now
are geographers learning the quantitative techniques. So it must
be asserted that it was impossible for geographers to answer the
type of question which once was common in school examination
papers. A question such as 'to what extent do geographical factors

account for the industry of Switzerland?' required the following type of answer:

Rock	$+ \ 0.0\%$
Relief	-10.0%
Climate	$+ \ 0.00073\%$
Soil	$+ \ 2.1\%$
Vegetation	$+ \ 5.36\%$

and the necessary data and techniques are still not available to answer such a question, even if it can ever be answered. Referring back to the question of what made geographers think they could solve the problem of whether or how man's actions are determined by his physical environment, one can only conclude that an intellectual arrogance combined with over-simplified data made the problem seem soluble. Nowadays more of us realise our limitations, and are aware that there are many things we still don't know about the world which are relevant to such a problem.

But of more immediate concern to geography student and teacher alike is the point that so much geographical effort during this century has been devoted to studying the effect of the physical environment on man. Have we reached the end of this trail? Is this the purpose of geography? Some geographers have made this their purpose, for example de la Blache[14] saw the purpose of geography as 'an attempt to see how far a certain kind of determinism is operative'. This 'certain kind' of determinism was nature controlling the extent of man's activities, not deciding what he must do. But many modern geographers would reject the suggestion that there is any element of determinism left in their work. The author believes the answer must lie between these extremes.

First, it is difficult to believe that geography could persist as an academic discipline with such a bizarre and biassed purpose as to study the effects of rock, relief, climate, soil and vegetation on man. This would put it on the same level as astrology. Two men stand out for their interest in this kind of study, Griffith Taylor for his interest in European settlement in different climates,[15] and Ellsworth Huntington for his interest in the influence of climate on the human mind.[16] There is room for such specialisation in geography, but for all geography to be directed to this end is another matter. If one agrees with the content, approach and questions of geography set out in earlier chapters of this book, then to concentrate on just the effects of the physical environment on man's economic activities would be to ignore so much more which is relevant to geography and of value to our knowledge as a whole.

Again, if one agrees with the earlier chapters, one might agree with the simple generalisation of geography studying the arrangements of the many different material phenomena on the earth's surface for the purpose of understanding the world. All the arrangements, distributions, connections and relationships can be studied of *any* two or more phenomena. Now geography which had as its purpose the understanding of one way cause and effect of physical phenomenon on human phenomenon, and excluded any consideration of human and non-material factors, would be partial and distorted geography indeed.

Geographers and teachers who confine themselves to 'geographic' factors are not just limiting their discipline but are distorting reality to the extent that their readers and pupils get a wrong explanation and are exposed to fallacious reasoning. In the location of a factory we say that there are physical, economic, political and psychological factors. There are still some geographers who would consider only the physical factors. Most geographers today would consider the physical and the economic factors. A few daring spirits are now admitting in textbooks what has long been known, that often political considerations override the physical and economic factors. Thus a firm may be forced to put its factory in an area of serious unemployment or well away from a dangerous border when all the other factors point to a better location. But one has yet to see in print, what interviews suggest, that an entrepreneur had to open his factory near London instead of Winsford in Cheshire or his wife would have left him; or a strange new factory was opened in Perthshire because the boss was keen on golf.

The majority of geographers confine themselves, in explaining the location of these factories, to the 'geographic' factors. Now they make it clear that they have confined themselves, but their readers, and especially schoolchildren, think they have the full explanation; while the geographer has sound reasons for not considering political, social and psychological factors, we have yet to see the growth of the discipline which does give us the other story. And how tedious! For the full explanation of the location of a factory one has to read not only a regional geography but also several social and economic histories. Much better to have the full explanation in one compact volume.

A less bizarre example should make the point clearer. Consider the full explanation of the functioning of a region to be like the full explanation of the functioning of a car engine. For the engine to run, petrol comes from the tank along pipes via pump and carburettor; electrical charges come from the battery along wires via the coil and distributor. When petrol vapour and spark get together in the cylinder, the engine fires. For the region to run, there need to be the influences of rock, relief, climate and soil on the one hand, and the influences of heredity, race, religion, education and basic animal instincts on the other. To me, it seems that the geographer who describes the region, and gives only half the explanation, by referring to the rock, relief, climate and soil, is like the idiot who describes a car engine and then explains its working only by reference to the petrol system and not the electrical. If the engineer justified his partial study by saying he was a hydraulics man and not an electrician we would just laugh. When the geographer refuses to attempt a full explanation of his phenomena we don't laugh. He says politics, finance, religion, psychology, morals, etc. are not his concern, and we accept it. The situation would be acceptable if another specialist were standing in the wings to take over where the geographer left off, but he isn't.

Houston[17] and others, however, see this as part of the nature of geography. For them, it is the study of certain phenomena on the earth's surface and a partial explanation of them. The phenomena have received adequate consideration in this book, but until now the partial nature of their explanation has not been emphasised. Just as one discipline cannot study everything in one place, neither can it give a complete explanation. The geographer gives a partial explanation with reference to factors such as physical conditions, position, distance, shape, size, economic conditions but in his own mind is clear that he does not consider other factors at all. This leads to the serious fault in teaching that very often the teacher is not aware of the factors which the geographer has ignored, so that finally, in the child's mind, the partial explanation seems to be the full explanation. Therefore one cannot be surprised when children and young people use faulty logic, have over-simplified ideas of cause and effect, and are completely unaware that factors besides the 'geographic' ones exist and operate.

Houston sees quite clearly that geography, like any other discipline, is no more than a partial understanding of reality.

'Disciplines, whether in the sciences or humanities, have progressed because they realised their limitations, making no pretensions to an attainment of absolute reality; but much confusion has arisen in sociology . . . because it has assumed itself to be more than a partial understanding of social phenomena.'[18] One must accept this point, and so, for the time being, the conflict remains, and one must conclude that some geographers believe that geography gives only a partial understanding. The rest of the full explanation is not often written down. If geographers can forget they have only part of the truth, then teachers certainly can. Children have not the intellectual concept of the nature of geography and therefore are likely to think they have full understanding and by the very nature of the work are liable to slip into determinist ways of thinking.

When such geography gives only a partial explanation it limits itself to some of the facts connected with the landscape and avoids trying to cope with total reality. Thus if a full explanation could be given by an ethnologist and a sociologist, the geographer confines himself to that part covered by the ethnologist, and ignores social, racial and psychological factors. The cultural landscape is very much what Kant meant to cover as 'physical geography': rock, relief, soil, vegetation, fields, roads, houses, factories, crops, animals, mines and people, i.e. all the visible, concrete, tangible things on the earth's surface. The intangibles and imponderables are left out. The term cultural landscape is American, and means precisely what we see when we go outside and look round. It is cultural in the sense that wherever we look the natural, physical landscape has been modified, used and built on by man. Many geographers realise it is impossible to re-create the original, virgin, physical landscape, as many others have tried, so instead of starting with rock, relief, climate, soil and natural vegetation they start with the land as it is now.

This, of course, is the most sound method for teaching. We pay lip service to starting from the known, the particular, and the familiar in teaching, and then go off and start teaching geography by trying to re-create a mythical, virgin, natural environment which might have existed 10,000 years ago. Start with the cultural landscape in which we live today, with the relief, shops, houses, parks, buses and climate that we know.

The approach through the cultural landscape has several advantages: it puts educational theory into practice, it studies first the effect man has had on the land (some of which is self-evident), and it limits geography to manageable proportions, and, through Sauer, wins increased respect for the discipline in the USA by establishing it as the study of land.[19] In contrast, by working up from rock, through all the abstract re-creations of pure relief, soil, vegetation cover, etc., and then going on to man and his works we create three problems: we are led into deterministic studies of the effect of land on man (not self-evident); we mistakenly think that geography has to be the complete study of man; and we are faced therefore with a task too large to tackle which brings geography into disrepute.[20]

Hartshorne argues that if we divide geography into physical and human phenomena we make the rest of our work illogical. Given this rift (which he thinks is the result of determinist thinking) then repeatedly we are drawn into studying the effect of man on land, the effect of land on man and then the fallacy of studying only the physical factors in man's activities and not the psychological factors. The two faults which are so distressing are, firstly, that we tend, again and again, to compare one physical phenomenon with one human, and much more rarely physical with physical, human with human. Yet geography is the study of the inter-relationships of many phenomena on the earth's surface, and not the study of the interrelationships *only* between physical and human. Secondly, given land and man, and having man as our main interest, then as geographers we feel bound to consider only the 'geographic' factors and to ignore the rest. Thus the division into physical and human is the cause of geography being only the partial study which Houston accepts. Yet Hartshorne writes: 'In fact, all geographers, whatever beliefs they may assert, recognise that we could not possibly explain human choices and actions solely in terms of relationships with the natural environment.'[21]

If one sees geography as the study of land and man, then one must do three things. One, study the effect of land on man. Two, make clear at all times that this is only part of the full explanation of man's activities. Three, measure precisely the extent to which the environment has affected man. Given this view of geography then determinism is an essential part of the study[22] and it is incumbent on the geographer to give a precise numerical value to the degree, proportion or percentage of determinism.

If, as Hartshorne suggests,[23] one sees geography as a study of man in his environment, one can study the interactions of any phenomena, can attempt a complete explanation, and can write in qualitative as well as quantitative terms. By environment in this case we mean man's actual environment, the cultural landscape which is the resultant of thousands of years of interaction between man and land. In this case the environment is rocks and houses, relief and roads, climate and clothing, soil and greenhouses, vegetation and motor cars, water and factories, trees and lamp-posts. One can study, say, the causality, connections and coincidence between the type of farming in an area and the total environment without bothering whether the present soil and vegetation are pure, virgin and natural, or changed by centuries of man's interference. Moreover, the environment will include the subsidies of the milk marketing board, the state of the road to the creamery and that fact that most of the young people are moving from the land to London.

For geography to have any value, the rift between physical and human phenomena must disappear. If, as Hartshorne believes, determinism is the cause of the rift, then we must stop thinking along deterministic lines, stop using deterministic phrases, even when we don't mean precisely what we seem to say[24] and so remove this fallacy of seeming to study the relationship of man and his natural environment. The easiest and safest way to avoid these errors is to start with the landscape as it is this morning, and to study the connections between those features we believe to be significant, even if they happen to be three or four 'human' features.

While Hartshorne would argue for the removal of determinist work and phraseology, and the present author would prefer it to be used with extreme caution, Martin[25] and Lewthwaite[26] insist that it is a necessary element of all geography. The following paraphrases of their papers are inevitably combined with other comments and the reader is urged to study the originals both for their precise reasoning and wealth of detail.

The burden of Martin's argument is that determinism is much more complex than most geographers seem to have realised. Matter, the physical environment, must have its effect on men's minds and ultimately on their actions. The difficulties lie in observing the middle condition where matter is affecting mind, and in isolating single cause and single effect. Martin puts the problem in its wider context, usually ignored in geography, that the whole state of the universe at one time is the cause and the whole state of the universe at a later time is the effect. He says that each discipline is defined by the way it looks at part of this total cause and effect complex. This is very similar to Houston's point that geography, sociology, etc., are only partial studies and explain only a few of the causes. To quote Martin verbatim: 'At the scientific level we can then try and state the laws according to which each particular part-cause takes its shape in the effects of the whole causal complex.' Then, 'geographers . . . however, cannot, like the experimental sciences, arrive at quantitative mathematical expression of their laws, owing to their inability to isolate one partial cause at a time in experiments'.

On the necessity for determinism Martin says that some geographers have argued that there is no such thing because the existence of determinism would demand that the same cause must be followed by the same effect. Geographers think they are able to show that the same cause is not followed by the same effect when they show two 'identical' natural regions with different types of land use. Martin insists that it is just the over-simplification of such geographers that makes two environments seem to be identical. The regularities which we *do* observe in the real world, however, could not happen if there were no determinism at all, so the conclusion Martin reaches, completely convincingly, is that determinism must exist, but the chain of cause and effect is very complicated indeed.

No two places, no two times, no two groups of people are exactly alike. In any place, at any time the environment has been one of the factors in determining what man will do. In Martin's view geography is the discipline which studies this set of factors. But because of the infinite variation of the factors and elements involved, the result of this deterministic cause and effect is unique in every case. 'Geographers do not assert that the geographic environment is the only *or even the most important* determinant of human activity, they merely state that their particular business is to examine this group of determinants rather than others.'[27] In agreeing wholeheartedly with Martin, the author would emphasise two points. First, the phrase too easily overlooked, that the influences which geographers study may be much less important in deciding human actions than other influences outside geography's terms of reference. Second, the important lesson of this paper seems not that determinism is inevitable, but that its processes are so complex. When the author seems to be a frantic anti-determinist, it is when he is in despair at the fantastic over-simplifications of geographers, teachers and students.

Lewthwaite echoes some of the points made by Martin, but his paper is most useful in clarifying the thoughts of those people who think they can observe environmental determinism operative in one place but not in another. Martin's paper convinces one that determinism exists everywhere, but one's knowledge of the world, provided largely by geography, inclines some to believe that while the few remaining Bushmen and Eskimoes may well be entirely controlled by their environment, Europeans and Americans must be entirely free. Lewthwaite puts this apparent contradiction in perspective, and ends the argument about the existence of determinism once and for all.

Lewthwaite starts an important part of his argument by saying that while general determinism may well be true, geographic or environmental determinism, asserting that environment is the single or most important determinant in man's actions, is extremely doubtful. He then demonstrates concisely that no geographer has been a complete determinist. Even Semple admitted that there are other factors but then chose to ignore them while riding her hobby horse. Griffith Taylor saw it as the geographer's job, however, to focus attention on the environment so that mankind could adapt to what he thought were the most important factors in our existence. Lewthwaite is not so clear about the other extreme, but he mentions possibilism and acknowledges that no geographer would suggest that man is completely independent of the earth's surface. At this extreme man may be able to decide his own actions from a wide range of possibilities, but the environment sets limits to these possibilities and also to the time and cost of achieving certain ends.

Lewthwaite's most valuable point is that we can use the concepts of determinism as models to get at the truth. Determinism and possibilism are the two ends of the spectrum, not contradictions, and several other models can be set up at stages along the spectrum to generalise about the working of environmental determinism at different times, in different places on different people. Here we see a geographer suggesting the use of models as a means to an end, and a very important end. He writes: 'The appropriate query to be placed against a conclusion is not "is this possibilism?" or "is this determinism?" but "is this true?".'

Febvre[28] was vehemently opposed to determinism, and he and

Brunhes, de la Blache, Bowman and Sauer have developed the idea of possibilism. An important point for Febvre was that man is not passive, 'man is a geographical agent and not the least'.[29] The possibilists were aware of the interaction of all phenomena and thus the unpredictable, active role of man in his environment, but they are aware that the environment does set some limits and exert some influences. Thus they are geographers, aware of the influences and limits of the environment, not fanatics setting out to prove a theory and distorting geography in the process. The possibilists see man as circumscribed rather than determined, and nature as permissive and an 'adviser' rather than a ruler.

Determinism has been revived in the modified form of stop-and-go determinism. In the preface to *Mainsprings of Civilisation* published in 1945 by John Wiley, Ellsworth Huntington states 'this book is an attempt to analyse the role of biological inheritance and physical environment in influencing the course of history'.[30] The sub-title of *Australia* by Griffith Taylor,[31] published in 1940, is 'a study of warm environments and their effect on British settlement'. Taylor, in fact, is the founder of stop-and-go determinism and his interest in the effect of climate on man is clear in most of his books.

This modified determinism, which takes account of the arguments which showed the errors of the old determinism, claims that nature has a plan. Because it seems that man is not controlled by his environment in many places this is explained away by the provision in the stop-and-go theory for man to follow the plan quickly, slowly, or even stop for a time. Moreover, as man cannot always see the plan, and might conflict with it, Taylor states that it is the function of the geographer to reveal nature's plan to the rest of mankind. Presumably, if there were no geographers to give the orders, man would not be determined. (The answer to this would be along the lines that the provision of guiding geographers is also part of the plan.)

This theory covers changes in land use like the change from arable to dairy farming in Denmark. If the change is successful, if any of man's efforts 'against' nature are successful, then this means only that man is fitting in better with the plan than he did before.

However much one dislikes the idea of a plan, much of what Taylor says makes sense. In many parts of the world man is up against high relief, drought, flood, heat, cold, poor soil or dense vegetation. Some types of land use would be much more appropriate than others, and the experienced economic geographer can suggest these. But from his writings and his footnote in *Geography in the Twentieth Century*[32] it seems that Taylor and nature would always advocate the easiest and least costly choice for man. Man should take the line of least resistance and, say, settle sparsely in a dry area, keeping cattle, rather than go to enormous trouble and great expense to irrigate the land, have a closely settled country-side rich with vineyards and market gardens. The stop-and-go determinist admits that several courses are possible, but that one is determined by nature as best in a given set of circumstances. This best usually turns out to be the obvious, easy, cheap line of least effort which obstinate man so seldom takes. Often man has goals, dreams of wealth and ease, and the price he is prepared to pay is seen to be prolonged hard work and poverty to achieve the goal.

Two other 'isms' deserve mention but have not gained much ground. Writing of human ecology Bryan[33] states: 'It is possible to study the way in which specific human groups or communities have adjusted themselves to *and modified* the natural conditions in which they live.'[34] But human ecology has been interpreted in too many different ways. Spate put forward the concept of probabilism[35] in a study of determinism, but considering the number of 'isms' it is perhaps fortunate that it has not yet been taken up widely.

An interesting paradox which gives food for thought is that while many determinist writers, unable to 'prove' determinism in complex modern societies, held that primitive people were controlled by their environments much more than advanced people, Brunhes and Kirchoff have insisted that sophisticated, technically advanced societies are much more dependent on their physical environments. These two conflicting ideas still have value in geography. By bearing them in mind during unbiassed regional work one may study the strength of the influence of the environment among other things, and be prepared to find that a manufacturing economy has had to fit into the environment much more carefully than, say, fishing, in order to make a profit.

From the two papers of Lewthwaite and Martin we have the following points on the matter of determinism. The geographer's concern with environmental determinism is just one aspect of a much wider problem. Man is inevitably influenced by his environment but nowhere completely and nowhere is he entirely free.

There can be no argument as to whether man is determined or not. The geographer, among other things, can evaluate only the degree to which the physical features have a direct effect on man's activities. The amount of influence is infinitely variable and Lewthwaite's idea of a spectrum or scale against which to match a particular example emphasises that we are dealing with differences in degree and not in kind.[36]

Answering the rest of the five questions posed above (p. 102), one must first of all insist that the examination of determinism must remain a part of the function of geography. However, one can understand those geographers who are tired of the determinist-possibilist controversy and therefore reject this as part of their work. The effect of ideas of environmental determinism has been to distort some geography so badly as to bring it into disrepute for bias, superficiality and fallacious reasoning. One would be dismayed to find that the sole purpose of geography is to examine the extent of the influence of physical features on man or even man's adaptation to his environment. This would be a very specialised purpose of a very odd discipline. But one must insist that these topics, together with man's use of the physical world, must be a part of geography, along with all the other functions and purposes mentioned so far.

1. Eyre, S. R., 'Determinism and the Ecological Approach to Geography', *Geography*, vol. XLIX, pt. 4, 1964, p. 369. See also Davies, W. K. D., 'Theory, Science and Geography', *Tijdschrift voor Economische und Sociale Geografie*, vol. 57, no. 4, July 1966, p. 125
2. Lukermann, F., 'Geography as a formal intellectual discipline, and the way in which it contributes to human knowledge', *The Canadian Geographer*, vol. VIII, no. 4, 1964, p. 167
3. UNESCO, *Source book for geography teaching*, Longmans, 1964, p. 2
4. Hartshorne, R., *Perspective on the Nature of Geography*, Murray, 1959, p, 100
5. Houston, J. M., *A Social Geography of Europe*, Duckworth, 1953, p. 25
6. Bryan, P. W., *Man's Adaptation of Nature*, ULP, 1933, p. 1
7. Houston, op. cit., p. 22
8. This we call physical and human now
9. Hartshorne, op. cit., p. 68
10. Buckle, H. T., *Civilisation in England*, 1857
11. Hartshorne, op. cit., p. 55
12. Semple, E. C., *Influences of Geographic Environment*, Henry Holt, New York; Constable, London, 1911, p. 2

13. Brigham, A. P., 'A Quarter Century in Geography', *Journal of Geography*, 1922, p. 13
14. de la Blache, V., *Principles of Human Geography*, Constable, 1926, p. 23
15. Taylor, G., *Australia: a study of warm environments and their effect on British settlement*, Methuen, 1951
16. Huntingdon, E., *Mainsprings of Civilisation*, John Wiley, 1945
17. Houston, op. cit., p. 53–55
18. ibid., p. 36
19. Hartshorne, op. cit., p. 58
20. Houston, op. cit., p. 25
21. Hartshorne, op. cit., p. 58
22. Martin, A. F., 'The Necessity for Determinism', *TIBG*, no. 17, 1951, p. 1
23. Hartshorne, op. cit., p. 59
24. Wooldridge, S. W., and East, W. G., *The Spirit and Purpose of Geography*, HUL, 1951, p. 33
25. Martin, op. cit., p. 1
26. Lewthwaite, G. R., 'Environmentalism and Determinism: a search for clarification', *AAAG*, vol. 56, 1966, p. 1
27. My italics
28. Febvre, L., *A Geographical Introduction to History*, Routledge and Kegan Paul, 1925
29. ibid., p. 63
30. Huntington, op. cit., and Preface, p. 7
31. Taylor, op. cit.
32. Taylor, G., Ed., *Geography in the Twentieth Century*, Methuen, 1951, pp. 161–2 fn.
33. Bryan, op. cit., p. 8
34. My italics.
35. Spate, O. H. K., 'Toynbee and Huntington: A study of determinism', *Geographical Journal*, CXVIII, 1952
36. Davies, W. K. D., op. cit., makes the same point as Lewthwaite

LAWS, DESCRIPTIONS AND

EXPLANATIONS IN GEOGRAPHY

Some authorities state that one change in geography is its progress
from the descriptive, through the classificatory, to the law-making
stage.[1] Certainly there is greater emphasis on classification and
law-making these days and this raises three points relevant to the
investigation of the nature of geography. One needs to examine
the kind of laws in question; the fact that some explanations
offered in geography might more properly be called description;
and whether geographers explain causes or associations.

There is still enough over-simplified deterministic phrasing in
geography for it to be necessary to distinguish between a law which
compels man to do something and a law which describes the way
in which he behaves. Further, with the increasing use of statistical
techniques, it is necessary to distinguish between scientific and
stochastic laws. The first distinction can be cleared up with a
simple example. When a geographer generalises that in cold
climates man lives by hunting and herding animals he does not
mean that some natural law forces man to live thus, but that man's
behaviour conforms to this law. In a similar, yet more complex
way, when Christaller stated that settlements are arranged in a
hexagonal pattern on a K3, K4 or K7 network, these mathematical
laws describe what is observed, and do not imply some materialist
mechanism which makes man arrange his settlements thus.

The reader may be familiar with the experience of hearing a
science teacher explain, say, the law of gravity, and then being
required to 'prove' the law for himself, experimentally in the
laboratory, with pendulum and stopwatch. Now a scientific law

of this kind can never be proved, one can only go on testing it *ad infinitum*. In a million experiments, if the law holds 999,999 times but does not hold on the millionth occasion, then the scientific law has been disproved and a new law must be formulated to fit the observed facts. For example, Einstein's observations disproved Newton's law of gravity and Einstein had to establish and test a new law which would fit the observed movements of the planets, stars and atoms.[2] But this century, natural scientists and sociologists have become aware that, while they can formulate general laws to which the mass of molecules or people conform, the behaviour of any one molecule or person seems to be completely random. Thus we have the idea of stochastic laws which say what the majority of the phenomena in question will probably do.

Lewthwaite mentions the confusion of these types of laws in connection with determinism.[3] If one describes the operation of determinism in terms of scientific laws, then one exception to the law disproves the whole concept. He describes the Russians as believers in economic rather than physical determinism, because, thinking in terms of scientific laws, 'the moment any influence other than natural environment is acknowledged . . . environmental determinism . . . has ceased to exist'. However, if we describe the operation of physical determinism in terms of stochastic laws, one exception to the general pattern of things by no means disproves the whole rule. Later Lewthwaite writes, 'it is no accident that deterministic systems and quantification have advanced together in the field of geography'. The statistical techniques in quantitative geography help us to formulate stochastic laws, but that is all. Robinson, Lindberg and Brinkman say the core of geography lies 'not so much in the search for environmental controls as in the search for general patterns of areal covariation and for local departures from these overall patterns'.[4] Thus models and laws are theories to be modified in the light of experience. Fifteen years earlier than this article, however, Martin[5] made a special comment on scientific and stochastic laws. He insisted that human actions conform to laws no less rigid than those which apply to physical actions. For him, scientific and stochastic laws are not different, it is simply that we have not discovered all the complexities which would enable us to make the stochastic laws more accurate and turn them into scientific ones.

In their article 'On Laws in Geography' Golledge and Amadeo [6] offer some help to geographers not familiar with fitting laws to observed phenomena. Only some of their more general points will be noted here, and the reader is referred to the article for practical advice. Golledge and Amadeo pose some very interesting questions about laws in general, but give some reassurance to those who feel bewildered. For example, after posing the question 'is a statistical generalisation a law?' they point out that no one can say because there is no single generally accepted definition of a law. Then again, on the matter of one exception disproving a scientific or physical law, but causing no embarrassment to a stochastic law, 'it becomes relevant to raise the question about the number of exceptions that may exist before a law becomes a no-law. This particular question has no absolute answer'.

Golledge and Amadeo go into some detail about inference (induction and deduction), probabilistic inference, cross section laws, equilibrium laws, historical laws, developmental laws, statistical laws, and composition rules. Their comments on induction stress the fact that human beings, when thinking, do not plod pedantically through all the stages of X, Y and Z. Our thinking is much quicker and direct than this, but these examples of formal logic show us how to fit the laws to observed phenomena, test our reasoning and logic, and set out the final presentation of our work in a clear and logical manner.

They admit that rather than proceed by induction, examining many cases before formulating a law, we grasp the law intuitively. The only danger lies in then not testing the law against many more cases. If we do test the law and it seems to hold up, the intuitive leap is called a brainwave; if we do not test it, it is called jumping to conclusions. Davies [7] makes exactly the same point, and adds that he thinks induction never works in practice. In the case of composition rules we find that different rules apply at different levels, from local to international, from village to city. This seems to have some bearing on the point which Haggett makes about different types of explanation being necessary at different levels. [8]

Golledge and Amadeo point out that most of the types of law which they describe allow prediction, in the sense that laws based on observations here and now allow us to make fairly safe statements about the future, about other places which we have not seen, and about the past, which we missed. Prediction may sound odd in this respect, but this, of course, is what makes the historical geographer's work possible and valid because phenomena obeyed the laws in the past. However, the cross-section laws are exceptions, because by their nature they can apply only at one time, under a given set of conditions.

Some of the remarks of Golledge and Amadeo would make more impact if they emphasised, as Martin does, that geography is not an experimental science. This fact causes difficulties in two ways. First, when a geographer has formulated a law or made a model, which he must test, he cannot set up controlled experiments in the laboratory. The only testing possible is to go out into the field to look for more of the phenomena and more of the systems in question. The grave danger here is in the temptation, or sub-conscious tendency, to look for those phenomena or systems which will sustain the law or model. One cannot examine all the cases in the world, so one must examine a large random sample in order to avoid bias.

Second, and equally important, so many of these laws depend on the conditions 'all other things being equal' and 'under normal conditions'. Now both on theoretical grounds from Martin, and from prolonged experience in the field, many geographers would never expect to find these conditions. All other things never are perfectly the same on the face of this planet and we would need to be omniscient to say what normal conditions are. Perhaps the most familiar, because long-standing, example of the necessity of all things being equal for a law formulated in one place to apply to another is the attempt to fit Christaller's central place theory, worked out in southern Germany, to other parts of the world. It does not fit without many exceptions, distortions, ifs, ands and buts, weightings and modifications. But it fits well enough for one to see its value, and to wish for something a bit more flexible. Many geographers see such laws and models not as descriptions of reality, of how things are arranged and behave, but just as ideal examples or standards against which one can compare the much more complex and irregular phenomena found in the field. The physicist, in the laboratory, can set up an experiment to test a law where he carefully arranges that the conditions for the experiment are the same each time. The geographer cannot do this, he deals with the sum total of phenomena actually working as a complete system on the earth's surface. The nature of geography therefore must be different from the nature of the experimental natural sciences because of the way the geographer has to work with the subject-matter he has chosen to study.

In his article 'Certain underpinnings of our arguments in human geography' Clark raises another vital point while considering determinism.[9] He, too, is concerned about the physical-human dualism in geography, and shows not only that the operation of determinism is much more complicated than some geographers seem to imagine, but also that it is very difficult to isolate causes. On the first point he writes: 'The human reaction to external stimulus, because it must pass through the mind, is determined not only by the nature of the external stimulus but also by the nature of the mind through which it passes.' Obviously, those geographers who confine their explanations entirely to physical stimuli are wasting their time. Clark put into words what modern geographers now take for granted, that physical conditions supply the medium in which human causal factors operate.

On the second point Clark states what some geographers will still not accept, that even a perfect correlation between two phenomena does not prove that one causes the other. The early, crude determinists seemed to reason along the following lines: there are physical and human phenomena and while physical conditions are relatively stable, human behaviour can change more rapidly. Therefore man has adapted to his environment and the physical conditions cause man to do what he does. There are many fallacies here, not the least being the assumption that geographers must juxtapose the physical and human. Clark insists that the last part of this argument could be rephrased. Man's aims, he argues, change so that he must use the environment in a different way and therefore in one place physical causes may be dominant, in another human causes may be dominant. He gives the obvious example that no physical conditions of any kind can 'cause' a padi field; but certain minds have reacted to the problem of living in certain environments by thinking up padi fields. Thus, states Clark, the word 'cause' can have two meanings: A compels B, and where A exists, B usually follows.

The present author would insist that the geographer is usually concerned with cause in the second sense—for two reasons. First, because the human agency, explained by Martin and Clark, is necessary for irrigation to be connected with deserts, and padi fields to be connected with south and east Asia.[10] Second, because 'where A, there, usually, is B' is just the kind of correlation demanded by the geographic approach.

When explanations of the causes are given for the phenomena which geography describes, different types of cause may be involved, and misunderstanding may result. For example, in the explanation of economic activity in a given region, it may be wrong for the geographer to assume that the people have always acted perfectly rationally, always worked hard and achieved the maximum possible, that their aims are the same as European aims and that they are motivated only by economic considerations. Contemplation of the farcical situation in Europe where every country must have an airline, and buy bigger and more expensive aeroplanes to outdo its rivals, when there is not the passenger demand, will show that this is not the case.

Geographers may also miss the cause of some distribution or location if they look only to other phenomena which have some areal or spatial relationship with it. The root cause or explanation may lie in history or in some intangible factor such as economic or political policy. Most geographers, of course, are aware of this, but the mistake is still made when some immediate cause is obvious and the indirect cause is not. If we take an individual farm or factory and ask the question 'Why is it here?' the answers we might get, 'Because the founder lived here', and 'Because this is the way my father farmed', are not the kinds of thing the geographer wants to know.

It begins to appear that the geographer does not really want to know the cause of things. To take the most familiar example, when we explain the location of industry we do not explain why each factory came into being; what we do is explain how this industry is situated in relation to other relevant features. There is no need to become historians and find out about Arkwright, Roe, Lever, Morris and others in order to explain why an industry started. As geographers we explain how it *is* located in relation to relief, population, communications, markets, raw materials, fuel and so on. We are explaining the spatial relationships as they exist now, not cause and effect in time. The correct type of geographic explanation is not that coal and iron ore were found in this region so John Smith started a blast furnace, but that the steelworks in this region are located thus, and thus, in relation to the coal, ore, limestone, railways, markets and people. Geographers explain how things are spatially related on the earth's surface, not necessarily what caused them or how they came about. The philosophers and historians can attempt that.

Wooldridge and East[11] see this distinction as only a matter of words; that we seem to talk about raw materials causing industry and physical conditions deciding the type of farming when really we mean that men started and decided these things and we are only describing how materials and industry or soils and crops are now related to each other. But this is more than just a matter of words. It seems that such slipshod phraseology has been used that students of geography first take the literal meaning of the words, misunderstand the purpose of geography, and then perpetuate this muddled thinking and writing. Wooldridge and East are too sanguine about saying one thing when one really means something else. Decide what you mean, then say what you mean.

Haggett,[12] in contrast, thinks these two types of explanation are a matter of scale. Thus at one level we may be concerned with individual decisions of where to start a factory or what to grow in a certain place. At a much more general level we are concerned to show how thousands of decisions have all tended to the same conclusion and there are broad patterns of land use and large areas of industry, the distributions of which can be explained by reference to other phenomena on the surface and to economic, social and political aims of the people in question.

If one studies the population of a large area, say Scandinavia, then as a geographer one must describe and explain the distribution and density, and the connections with the other surface phenomena. As it is impossible to ask every person why he or she lives in each particular place, the type of explanation the geographer gives is that the mass of the people live in these numbers in these places because the relief, climate, soil are thus, and the opportunities for farming, mining, manufacturing are thus. This explains the distribution and density of population in relation to physical features and land use, and no more. The reasons for individuals being there are because they were born there—they went there for better jobs—because they cannot get out, and so on. The distributions in relation to physical features and land use are what the geographer wants to know. But one must insist that he then refrains from saying that physical features and land use cause or even explain the distribution, even if this is only the result of his cheerfully loose use of words. All the geographer can say is that the phenomena are connected spatially in the way he has described. If the patterns are repeated elsewhere, then a law can be formulated to describe this regularity of connection, but the laws describe how phenomena are connected and how they do work—not how they must be connected and work.

Among other things, reasoning along these lines will help to avoid two traps. The first is that of the most superficial determinism, stating that a certain physical condition results directly in a certain human response. The second is that of not being able to account for the exception to the rule. Stochastic laws, used to describe the characteristics of the phenomena in question, leave one open to be able to deal with, say, the case of very dense population in an unfavourable area on its own merits.

To sum up, one change in the geographical approach illustrates this matter very well. At one time geographers tended to try to explain the site of every individual settlement. Nowadays, they explain the distribution of villages, towns and cities not by reference to unique sites, but by reference to the spacing of the settlements, the size of area needed to support them and the size of area each serves. Now the people of each village and town had to choose a particular site in the beginning. This historical choice is unalterable and explains the site of each place. But the laws of central place theory explain how settlements are distributed in space when there is no distortion due to raw materials and break of bulk points. The distributions of settlements conform to these laws, or, more properly, the laws fit the observed facts tolerably well. But the laws did not force the Anglo-Saxons, Danes, and the rest to settle in any particular place. The law describes and explains how they did behave, not how they had to.

Both Martin and Lewthwaite, in stressing the long, complicated and often indirect chain of events between causes and effects, imply that geographers would be taking on a superhuman task if their purpose was to study the determinants of human actions. Geographers may, in fact, have worded their questions too vaguely to get the most useful answers. Knowing that Wooldridge and East[13] openly admit that geographers write one thing when they mean another, it is suggested in all seriousness that when some geographers have stated that they are looking for the laws which determine human behaviour, they have in fact been looking for laws which describe and explain human behaviour. Certainly they would have more chance of success with the latter quest.

Modern research, providing data which can be analysed using statistical techniques and computers, shows that in the mass human beings behave in 'lawful' ways. By this we mean not necessarily that they must behave this way, but that they do behave this way. Laws are formulated to describe the ways in which

masses of people do behave; each individual may behave errati-
cally, there are exceptions to the rules, which are therefore stochastic laws. Moreover, there is the implication in many papers which
describe human actions in these terms that if the behaviour can be
described by a mathematical formula, then this formula is also an
explanation of the behaviour. In other words, if most people
behave in a regular and predictable way, then no other explanation
is necessary. For example, in homogeneous farming areas a law
has been formulated to describe the rank and size of towns.[14] If
the towns are ranked by size of population, and rank is plotted
against size on double-log graph paper, the result is a straight
line. The result is at once impressive and frightening—that all the
people, making millions of separate decisions about where they
will live, have behaved in this neat and predictable way.

There is not necessarily any physical cause and human response
involved here. Each individual makes a rational decision, some
individuals behave in irrational ways, but these millions of
decisions and actions result in average behaviour. The purpose of
the geographer is not to be concerned with the individual decisions
which decided the type of farming, which started the mines,
located the factories and towns. His purpose is to study the resulting distributions, patterns, locations and associations in space,
using laws which describe the average results, the patterns, the
ways in which people do behave on the earth's surface. As Davies
says,[15] geographers may not be able to ascertain the causes, but
they can formulate laws describing the actions of the majority of
people.

In using stochastic laws, however, geographers could again be
diverted from the straight and narrow path. Processes, the working of systems, can be described by laws. But it has been argued
earlier that the study of processes and systems is not a geographical approach. The geographer must study the distribution,
arrangement and connections of phenomena on the earth's surface, and there is not sufficient evidence yet for one to be able to
assert that distributions and spatial arrangements *on that irregular
surface* can be described by laws. Obviously, on the earth's surface,
the distribution of rocks and relief is unique. One can discern
recurring patterns in climate, vegetation and soil, but these are
vastly distorted by the rocks, relief and shapes of the continents.
One would welcome some statistical technique which could decide
whether the patterns or the exceptions were more significant. If

these physical phenomena show insufficiently regular distribution, arrangement and connection, then one would expect human phenomena to show even more irregularities—but only research and analysis will show this.

Davies argues that the nature of geography must change as the world, our ideas and our methods change. Later he says that the main change in geography is that laws put forward vaguely in the past can now be formulated precisely, tested rigorously, and re-formulated as necessary. The techniques are there, but the present author still doubts the last stage in the process, knowing that human nature has not changed. There are those who try to make reality fit the models. However, Davies insists that for geography to progress, geographers must establish laws or models about such things as how two or more phenomena combine, the distribution of human activities, optimum population distribution and land use, how phenomena combine in type-regions. He states that we need a theory of how phenomena combine in space of different categories.

The age-old dilemma confronts us once again. In saying that laws and models must be created and tested, Davies would avoid one danger only to be exposed to another. One great danger in geography, as in any other discipline, is to go on collecting facts *ad nauseam* and never knowing what to do with them. The inductive process involves having a purpose, collecting facts, and formulating a generalisation. Collecting facts is only the middle, routine operation. So Davies' idea of making a law or model first seems to be an excellent way of avoiding collecting facts for the sake of collecting facts. But there is just as great a danger in this process because then the overriding temptation is not to test and destroy one's beautiful law or model, but to prove it. Geography needs more than Davies' device to ensure that all its exponents produce scholarly work.

Finally, one must disagree with Davies when he says geography must change. The content may be added to, the techniques must be improved, but one would argue that if the purposes change as well then geography does not exist. He seems to be in favour of the short-term expediency which Haggett and others deplore. He is in favour of a goal-directed or problem-oriented programme. Now, if Davies put forward his paper as a revolutionary idea for geographers, one cannot agree with him. If he, and others, want a discipline with completely different aims and purposes from the old-fashioned geography, then they must call it something else.

Neither the phenomena observed, nor the techniques employed, are unique to geography. If it stands as a separate discipline, then it is distinct because of its approach to the phenomena, studying their arrangement on the earth's surface, and because of its purposes, the kinds of things it wants to know. The study of the arrangement of phenomena on the earth's surface has always been implicit in geography. Even when such things as landscape, determinism or regional description were the main interest, site, location, place, space and distributions were always vital considerations. The only change here is that the geometry is in the limelight at the moment. In this respect, essential elements of the geographic approach have waxed and waned.

On the question of purpose, as distinct from approach, the evidence suggests that more and more men who call themselves geographers have an increasing variety of purposes. If this is growth and progress, and not a complete break from the past, then it is to be welcomed. Some personal conclusions about the geographical approach and the purposes of geography will be made in Chapter 11. To conclude this section, however, there are signs that the changes in geography are so great that it is difficult to see what the many contemporary specialist branches have in common, and difficult to identify present and past geography as one and the same discipline. If the number of branches and purposes increases, but one can find a continuous unifying theme, then the changes are acceptable. If purpose replaces purpose one after another, then geography must become disreputable. When some geographers reject subject-matter as irrelevant, because human phenomena are changing so rapidly that facts are not worth learning; when others scorn long established techniques; when the aim to understand the earth's surface is replaced by the aim to plan and change it, then one thinks that perhaps geography is not changing, that it is being replaced by something else.

1. Houston, J. M., *A Social Geography of Europe*, Duckworth, 1953, p. 18; Golledge, R., and Amadeo, D., 'On Laws in Geography', *AAAG*, vol. 58, 1968, p. 760
2. Einstein, A., and Infeld, L., *The Evolution of Physics*, Cambridge, 1938
3. Lewthwaite, G. R., 'Environmentalism and Determinism: a search for a clarification', *AAAG*, vol. 56, 1966, p. 2

E

4. Robinson, A. H., Lindberg, J. B., and Brinkman, L. W., 'A Correlation and Regression Analysis Applied to Rural Farm Population Densities in the Great Plains', *AAAG*, vol. 51, 1961, p. 211.

5. Martin, A. F., 'The Necessity for Determinism,' *TIBG*, no. 17, 1951

6. Golledge and Amadeo, op. cit. See also Salmon, Wesley, C., *Logic, Foundations of Philosophy*, Prentice Hall, 1967

7. Davies, W. K. D., 'Theory, Science and Geography', *Tijdschrift voor Economische und Sociale Geografie*, vol. 57, no. 4, July 1966, p. 125

8. Haggett, P., *Locational Analysis in Human Geography*, Arnold, 1965, p. 263

9. Clark, K. G. T., 'Certain underpinnings of our arguments in human geography', *TIBG*, no. 16, 1950, p. 13

10. The word 'connected' is used in the sense described in Chapter 7, (p. 75)

11. Wooldridge, S. W., and East, W. G., *The Spirit and Purpose of Geography*, HUL, 1951, p. 33

12. Haggett, op. cit., p. 8

13. Wooldridge and East, op. cit.

14. Cole, J. P., and King, C. A. M., *Quantitative Geography*, John Wiley, 1968, p. 483

15. Davies, op. cit., p. 126

I I

RECENT CHANGES IN GEOGRAPHY

In the light of the commentaries of Freeman[1] and Houston[2] it would be rash to assert that the changes to be described are either permanent or more than changes in emphasis. Yet there is a belief that geography is both changing and improving and one can only point to what seem to be the most important recent changes and await the test of time. Therefore the changes mentioned are put forward as a basis for discussion and to clear the ground for some comments on the nature of geography in the last chapter.

Three changes in the content of geography are worth mentioning. First, the variety of subject-matter is increasing, and we see both new general geographies such as geographies of religion and disease, and the inclusion of such phenomena in regional studies. But where specialists have pursued the study of the content rather than the study of the arrangement of that subject-matter on the surface of the earth, they have progressed beyond geography.

Second, paradoxically, in some directions we see the reduction of subject-matter, a decrease in the number of phenomena studied, because more and more geographers are putting the emphasis on connection. The old-style regional geographer was prepared to study all phenomena in a region, just because they were there, assuming some connection. With the greater interest in general geography now, the study is directed to those phenomena which prove to have causality and connection, so the content of both regional and general geography is narrowed down in this respect.

The third change is more subtle; a change in which aspect of the phenomenon is important. Once the texture and parent material of soil was considered vital, now it is the profile. How and why fishing and forestry are carried on is now of less interest than where. The physical conditions of farming now attract less attention than the layout of farms, the relationships of buildings to fields and roads, of farms to markets. Roads and railways still attract the geographer's attention, but less as to how they have to fit in with the relief of their routes, and much more as to how they connect towns and provide networks with better and better connectivity. Perhaps the change is most obvious in the geographic study of settlements. House types and building materials, with the emphasis on the crude, simple determinism of materials and climate, are now of less interest than the distribution, grouping and spatial arrangements of houses and settlements. Where town plans, unique sites, historical growth concerned the geographer in the past, he is now concerned with location, function and sphere of influence. The overall impression is that what were the main topics of geography a quarter of a century ago are now mere side issues. In the case of subject-matter this seems to be a decided improvement.

It would be equally logical to place the last paragraph under the heading of 'approach' rather than of 'content' and this fact stresses another change or trend. One of the greatest improvements in recent geography, in the author's opinion, is the increasing emphasis on the importance of the way the geographer approaches his subject-matter. To the geographer, the subject-matter is less important than the particular aspects of it which he examines.

The most obvious recent changes have been in method and technique. But too many different types of things are included under such blanket phrases, and some changes may not be the ones intended. The new quantitative techniques and general systems theory are new only in the sense that they have been used extensively in geography only in recent times. Moreover, they are new only in the sense that at best they enable the geographer to do more accurately and more quickly what he has always been trying to do. There are two dangers, however: first, that older but still useful techniques are discarded; second, and more important, that in welcoming with open arms all the quantitative techniques of the statistician, lock, stock and barrel, the geographer is trying to use some which are completely unsuited to his purpose and may lead him astray.

For example, one can observe the two extremes of former geographers processing data by methods which do not relate the phenomena to place or area in any way; and others moving on to spatial analysis which has lost connection with any real surface on this earth. However, once these extremes are recognised and avoided, the new techniques improve geography in several other ways as well as making the analysis quicker and more accurate.

The use of strict sampling techniques helps to avoid unconscious bias. The concept of statistical significance can now reveal to a geographer those avenues which are not worth following, where he would be wasting his time trying to rationalise, as well as revealing those which are worth while. Equally important, the older methods of induction which necessitated much routine plodding can now be replaced by the testing of hypothesis, law or model. Again, this idea is not new, it is just being consciously and rigorously applied in geography at last.

Along with all this goes the realisation that geographers are dealing with stochastic laws which, when formulated as accurately as possible, can describe how the majority of people do behave. Provided that enough samples have been taken, that the law has been tested against further examples and revised as necessary, then by its nature it can allow for the exceptions which are the result of the decisions of certain unusual individuals. Those geographers who have realised this have also realised that they are therefore not required, logically, to try to explain original or historic causes.

Here again we can witness one of the greatest improvements in recent geography. Fifty years ago geographers were so close to the industrial revolution, and had so little evidence of revolutionary changes in farming, that their explanations of the location of industry and the distribution of farming was inextricably mixed up with an explanation of why the factories were started and assumptions as to why farmers farm in certain ways. Now as we get further and further away in time from the original factories, and as farming under the same physical conditions continues to change as the result of economic factors, then more and more do we see clearly that the explanation of the start of these things is a matter for the historian. What the geographer explains is how the factories are now located in relation to other things. For example, the Pennsylvania steelworks were started using local coal and iron ore. Later they used iron ore from Lake Superior. But their present location is in relation to coal from the Midwest and iron ore from Labrador and Venezuela. The contemporary geographer does not need to mention the early nineteenth century, but he must explain the present location in relation to accumulated capital and skilled labour in Pittsburgh, the markets near the steelworks, and the present distribution of fuel and raw materials.

An important aspect of this type of general geography is that examination of particular cases of farming, mining, manufacturing and settlement in many parts of the world reveals both repeating patterns and logical arrangement. The fact of repeating patterns means that the geographer can make laws, models or generalisations, and the fact that man usually arranges his activities on the earth's surface in a logical way is sufficient explanation in itself. Thus our stochastic laws provide explanations of logical spatial behaviour on the earth's surface as well as description.

To sum up, the basic change here is from an examination of the reasons for many individual actions, which is not geography, to an examination of the present behaviour and arrangement on the earth's surface of the majority of people. The geographer is concerned with the present, with arrangements on the earth's surface, and with masses of people who behave in the same way. Hence the importance of the concepts of the median, mean, and especially the mode.

Two changes can be discerned in the practical approaches considered in Chapter 5. While specialisation in one branch of general geography may be necessary for a geographer to make progress, there is no greater significance in the division between physical geography and economic geography than there is between economic and social geography. While this artificial split between physical and human geography may be disappearing now, there is a danger that it will be replaced by a split just as serious. In the nineteenth century, physical geography, including the real landscape, was the core of the subject, and the relevance of human geography was in doubt. We are approaching the time when the relevance of the landscape itself may be in doubt and spatial analysis come to be regarded as the core of the subject. The study of landscapes and geometrical spatial analysis combine much less easily than do the studies of human and physical phenomena on the earth's surface.

In the matter of the other practical approach, that of taking a region rather than a topic, there is the attempt by some observers to assert a change for which the author can find no evidence. Haggett,[3] Saey[4] and others believe that there is a necessary historical trend in that regional geography is giving way, and should give way, to general geography. Having examined regional geography, the author has been mistaken for a defender of regional geography in this country and its assailant in the USA, and so hopes this is proof of objectivity. Haggett and Saey insist that the change from regional to general geography is a matter of logical progress. The author has no axe to grind here, but can find no evidence of the necessary corollary of such progress, namely that of general geographers making use of data available in regional geographies and regional geography being used to test the hypotheses put forward by general geographers. This may happen in the future, but one can only wait and see. At the moment both types of geography continue to flourish, with insufficient obvious exchange of facts and concepts between the two.

It would be much more encouraging to be able to state definitely that there were certain changes in the purpose of geography. But the wished-for changes are not clear, and one can therefore only pose three questions. Is geography moving away from the fallacy of believing in geographical features and geographical factors? Simons[5] insists that it is, but Honeybone[6] persists in using both these terms. Does a harbourmaster think he is running a geographical feature? Does an agronomist realise he is interfering with geographical features? There are no geographical features and no geographical factors. The geographer may study the effects of any factors on the geography of any features, and by the geography here we mean the distribution, location and spatial arrangements of the features in question. Or again, if the geographer asks how the geography of certain features or places affects such things as their development, he is referring to the distribution, site, location and internal arrangement.

Is geography still satisfied with the partial explanations on which Houston and Martin insist? If a geographer describes the distribution of a phenomenon or the spatial arrangements within a region, and then explains them only in terms of physical features, then this is a most incomplete explanation. In practice geographers have gone beyond this. The location of industry is explained in terms of labour, transport and markets in addition to sites, raw materials and fuel. The distribution of farming is explained in terms of economic factors as well as physical factors. Now, are the men who have written thus geographers, as defined by Houston[7] and Martin?[8] Have they gone too far? Or should geography be so defined as to include logically a full explanation of what is described?

Are geographers adding to their skills and purposes or exchanging one for another? In other words, is there cumulative progress or just haphazard change? Some observers state that description gave way to classification, which in turn is giving way to law making. Yet all three processes are important and could be used together. Similarly, quantitative techniques may be replacing fieldwork and mapwork, instead of augmenting them.

One thing is clear, that the interests, methods and purposes of the increasing number of geographers are changing rapidly. These interests and purposes may lead them into research which is at one and the same time infinitely worthwhile and yet different from what is commonly accepted as geography. If a discipline called geography is to exist as a continuously recognisable entity, then these men need to define new disciplines for themselves rather than distort geography to fit what they do. We see geomorphologists, spatial analysts, regional students and economic planners with increasingly different aims. To take one as an example, one function of geography, among many, would be made its sole purpose by a few practitioners. The author believes that the fact that geography is becoming of practical importance in social and economic planning is one of the best trends, but at the same time he would distinguish between academic geography as useful training and help for the planner, and a geography of which the sole purpose was to plan the future. Geography has a function in revealing practical problems and inefficient land use on the earth's surface. Its purpose should not be to solve these problems and improve the land use alone.

Thus we see that the same subject-matter is being analysed by the same techniques but with an ever increasing variety of approaches and purposes. Many of these changes, as mentioned, are improvements which can do nothing but improve the intellectual standing and satisfaction of geography itself. But some of the changes in approach and purpose are so extreme, so diverse, that one would suggest that it is high time some new disciplines were hived off from geography, before the work done by 'geographers' under the name of 'geography' becomes so diverse as to defy definition.

1. Freeman, T. W., *A Hundred Years of Geography*, Duckworth, 1961
2. Houston, J. M., *A Social Geography of Europe*, Duckworth, 1953
3. Haggett, P., *Locational Analysis in Human Geography*, Arnold, 1965, ch. 1
4. Saey, P., 'A New Orientation of Geography', *Bulletin de la Société Belge d'Etudes Géographiques*, tome 37, no. 1, 1968
5. Simons, M., 'What is a Geographic Factor?', *Geography*, vol. 51, no. 3, 1960, p. 210
6. Honeybone, R. C., 'Sample Studies', *The Geographical Association*, 1962, pp. 6–7
7. Houston, op. cit.
8. Martin, A. F., 'The Necessity for Determinism', *TIBG*, no. 17, 1951

I 2

ONE GEOGRAPHY OR MANY?

One aspect of the nature of geography encourages this subdivision and branching out and makes it appear to be quite normal. For a long time geography seems to have been divisible into a number of mutually exclusive branches, which on the surface have appeared to be quite logical.

Physical geography	*v.*	Human geography
Regional geography	*v.*	General geography
Historical geography	*v.*	Contemporary geography
Determinist geography	*v.*	Possibilist geography
The study of:		
Formal sites	*v.*	Functional locations[1]
Landscape	*v.*	Space

And all the separate branches which study, e.g., landforms, climate, land use, population.[2]

Geography seems to allow different workers to approach the subject-matter in different ways, and yet for them all to have something in common which makes them geographers and different from workers in any other discipline. Obviously not subject-matter or techniques, for these are common to many disciplines. Obviously not purpose, because this varies too much, so the approach of studying arrangements on the earth's surface is suggested as the common denominator.

Table 2 sets out the various branches of the subject. As these branches of general geography are also combined into regional

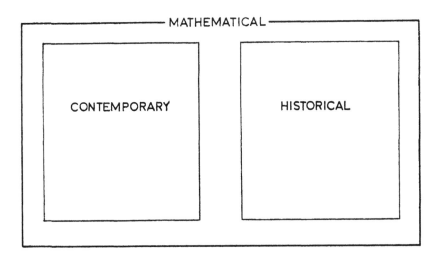

MATHEMATICAL	
CONTEMPORARY	HISTORICAL

CONTEMPORARY GEOGRAPHY							
SPECIAL GEOGRAPHY							
GENERAL GEOGRAPHY	1 North America	2 South America	3 Europe	4 Africa	5 Asia	6 Austral.	7 Antarc.
Landforms				░			
Climate				░			
Soils				░			
Plants				░			
Animals				░			
Economic				░			
Social				░			
A Urban				░			
B Settlement				░			
C Population	▨	▨	▨	▨	▨	▨	▨
D ?				░			
POLITICAL				░			

NOTES: D = Many new branches growing here
░ Special (regional) geography of Africa
▨ General geography of population which considers its regional variation throughout world

TABLE 2: GEOGRAPHY

geography, it can be seen that these are the two main aspects of the subject. However, there are other aspects, equally well known, such as historical geography, mathematical geography and cartography, which at first sight do not seem to fit into this plan.

Historical geography must first be distinguished from the history of geography and that of geographical exploration and discovery. The historical geographer does not chronicle the adventures of the explorers, neither does he describe what has been written in the past. The historical geographer writes regional or general accounts of the geography in past times. He attempts to do for certain stages in the past what the ordinary geographer does for the present; in effect, to write the geography of an area or a topic as if a modern geographer had been alive at the time in question. Ralph Brown[3] attempted to show what the United States were like at the time each region was settled. Darby[4] attempted to reconstruct England just after the Norman conquest as far as the Doomsday record permitted. East[5] has reconstructed the human geography of Europe at several stages, while other historical geographers have dealt with topics, rather than areas in the past.[6]

Therefore, historical geography does not fit within the framework of geography as we know it, but stands side by side with the geography of present times. The number of works on historical geography is not as great as the number on contemporary geography, but as historical geography embraces both general and regional works, and involves all the branches listed, it is not a part of geography, in the sense that social geography is a part, but a complete and separate entity.

The position of mathematical geography is at first even more difficult to assess. Dealing with the size and shape of the earth, latitude, longitude and time, it has affinities with geophysics. As it also deals with the earth's rotation, tilt, movement round the sun, and thus the seasons, it also has some connection with astronomy. Thus, it is vital at both ends and throughout the range of geography. However, by itself, mathematical geography reveals very little about different regions, or about the topics which form the content of geography. While it has a name, and a vital role, it seems more in the nature of a tool or a medium, just as the English language is a medium through which British and American geographers express some of their ideas.

Table 2 shows how the combinations of phenomena and parts of the earth's surface can give regional or general geography. The table also emphasises that historical geography must exist as separate from and equal to the geography of the present day. Regional and general, contemporary and historical geography are mutually exclusive and logically must exist as four divisions of the greater whole.

The studies of determinism and possibilism, or of formal sites (spatial connection) and functional location (spatial interaction) are not mutually exclusive. Lewthwaite[7] demonstrates that determinism and possibilism are two ends of a continuum, while the author has shown that form and function are two qualities of the same thing.[8]

However, one must deplore the artificial division into physical and human branches, which have no logical grounds but are the result of historical development. Eyre[9] makes practical suggestions for the healing of this split. In contrast, while the difference between physical and human specialisms is over-emphasised, it seems to the author that specialisation in the branches of general geography is both logical and a practical necessity.

However, a new dichotomy is threatened between the studies of real places on the earth's surface and of geometric space on paper. To avoid another century-long argument in geography similar to the physical-human and determinist-possibilist controversies it would seem better to insist on two distinct disciplines as soon as possible. These would be Topography, concerned with the real landscape and particular cases and places, contrasted with a new Geography which would be concerned with spatial analysis and generalisations.

Berry[10] has suggested that regional and general geography are not different approaches, but are just the two extremes of a continuum which he likens to a three-dimensional matrix. Similarly Lewthwaite has shown that determinism and possibilism are two extremes of a continuum. Berry, Lewthwaite and others are striving to keep the branches of geography together by demonstrating their logical interconnections. Haggett[11] has taken this a stage further when he states that many of our problems result from the failure to recognise geography as 'multivariate' in nature. He suggests that the solution is to restore the 'tripartite balance' of the Earth, Social and Geometrical sciences. It is in this context that the author considers Chapter 5 so important. The philosopher can have this ideal concept of geography in his mind; he can visualise Berry's three-dimensional matrix or Haggett's scientific trinity. But the geographer starting some research next Monday morning has to make a practical approach to part of the whole. *Both* the theoretical concept and the practical subdivision are essential.

Geography is multi-dimensional not only in the number of topics and regions of the world which can be included in one study, but in the approaches through any one of the branches of general geography; through historical or contemporary synthesis; through spatial connection or interaction. Geography is multivariate not only in its combination of natural science, social science and mathematics including geometry but also in the ways different geographers may combine these elements. Haggett is not alone in insisting on this combination of theory and practice, geometry and landscape. Hodder,[12] in his study of the tropics, demands that models and plans be fitted to the unique and individual regions. Putting emphasis on the geometrical element in this multivariate discipline, James[13] writes of geography: 'Its distinctive mission is to develop theory regarding space relations on the face of the earth and to describe the modifications of ideal sequences of events when these events occur in the presence of other kinds of events in the same area.' To translate the second half of the quotation from systems jargon, James is advocating both the formulation of spatial theories and their testing against real landscapes where phenomena interact, modifying and distorting the neat theories.

Several authorities remind the geographer both of the need for some theory, such as that provided by 'geometrical science', and of the infinite variations of the real landscape which makes the application of the theory so difficult. Martin stresses that while the theory of determinism is necessary each case of its working out is unique. Christaller's theory helps geographers merely to make some sense out of landscapes where no two are ever alike. Golledge and Amadeo, when urging that laws are essential to help us make sense of what we observe, still admit that the necessary condition of 'all other things being equal' never obtains. The point of all this is that if it is difficult to conceive of geography existing as one perfect whole while reading quietly in one's study, then the subdivisions mentioned above are certainly necessary for the working geographer tackling some particular topic. But what is also necessary is for him, from time to time, to check where he fits into geography as a whole.

The idea that there may exist several geographies rather than one is strengthened when one examines the definitions made by different people. The list of definitions below is not intended to be exhaustive, but is provided to show both the difficulty of a succinct definition and the variety of results obtained. According to the various authorities geography is the study of:

(1) Landscapes

(2) Places—James, Lukermann

(3) Space (in particular the earth's surface)—Kant

(4) The partial effects of natural environment on man—Houston, Martin

(5) Areal patterns of covariation—Robinson, Lindberg and Brinkman.

(6) Location, distribution, world-wide interdependence and interaction in place—Lukermann

(7) The combination of phenomena on the earth's surface

(8) The vast system of man and nature

(9) The man-earth system—Berry

(10) Relationships and reciprocity in the ecosystem—Morgan and Moss

(11) Human ecology

(12) Areal differentiation of interrelated phenomena on the earth's surface of significance to man—Hartshorne

These definitions fall into two groups; those which in fact define some part of geography, and those which attempt to define geography as a whole. The drawback with both types is that the shorter and more all-embracing they become, the more they need a long commentary to explain them to the student or layman. There is a need for something longer than these sentences to sum up the main points which are discussed at varying lengths in the books on the nature of geography. Certain points need to be emphasised before such a paragraph is attempted.

In one way or another the definitions stress the importance of the earth's surface, and particular places on the surface, in contrast to space in an abstract sense. Because of this, the geographer is concerned with the sum total of phenomena which exist in the world, but only just as they are arranged and combined on the earth's surface. The general geographer may concentrate his attention on just a few of these phenomena, such as climate, soil, farming and markets, but the geographic approach, as distinct from any other discipline which may be defined, is to examine how these things are arranged and logically combined on the earth's surface.

If this point is accepted, then it seems that there is a greater difference between regional and general geography than Berry would accept. Regional geography tries to examine all the important phenomena on the earth's surface, and therefore to deal with total, complex reality. Now when a general geographer selects landforms, soils, manufacturing or population for his main topic, to a greater or lesser degree he isolates this topic from the rest of the phenomena on the earth's surface. Hartshorne and Martin stress that geography is not an experimental science, and the more a general geographer isolates one phenomenon the nearer he gets to the idea of a natural scientist isolating some specimen in the laboratory. Conversely, he gets further away from the earth's surface and the real interconnection of phenomena in place.

However, even regional geographers now recoil from describing all the phenomena in one place, just because they are there, and tend to concentrate on those phenomena which they discover are interconnected. At a time when regional description is in a backwater, it may be necessary to conceive general geography, compage regional geography, and full descriptive regional geography as three quite separate branches. Compage geography will not include phenomena which are simply characteristic of a place unless they show some logical arrangement in space and connections with other important phenomena. Yet thinking of geography's wider function and obligation to educated laymen, as distinct from professional geographers, full, orderly regional description may still be required outside the profession.

The remarks about experimental science and the concern with the real earth surface bring up the point of samples and models. The full combination of phenomena all over the earth is so complex that both sampling and the construction of simpler models are essential. But some geographers seem to rest content with the production of a simple model which is a pale shadow of reality. If the everyday world and particular places are important to the geographer, then such models must be used to compare with a real place or with actual phenomena in place, and the final comment made on the world, not the model. Any study which ends up with an abstract concept of physical or human phenomena has become a natural or social science, and is no longer geography.

The sum total of phenomena mentioned above, obviously, includes man and all his activities on the earth's surface. In one way or another, the extent of the geographer's pertinent enquiries is limited not only to arrangements in space, but to those phenomena and spatial arrangements which are important to man in living and making a living. The criterion of significance to man helps the geographer to limit his work. Here the concepts of human ecology, the ecosystem, and the one man-earth system are important. In particular one would contrast the idea of reciprocity, emphasised by Morgan and Moss,[14] with the older idea of the one-sided and partial effect of physical conditions on man's activities. Modern geography studies the interactions, the reciprocal relationships of any phenomena in place on the earth's surface, not one-way actions of land on man.

Here we have the basic idea, inherent in ecology, ecosystems and geography, that each object, phenomenon or event in a place both affects, and is affected by, all the others. Thus the geographer, in particular, in studying the assembly of things on the earth's surface, must be aware that each affects the other. Man affects the land, vegetation affects the climate, industry affects the towns just as much as climate affects man. Morgan and Moss go into some detail of ecosystems and reciprocity, seeing them embracing much more than Eyre[15] suggested. 'A geography of living things is concerned ... with all forms of relationship affecting the distribution, location and space organisation of living things as they appear on the surface of the earth.' In contrast to older regional geography they insist on a kind of areal analysis rather than regional synthesis. They write of the Biocenosis or group of plants and animals, including man and his farming, and the Biotope or habitat in which they live, together making up the Ecosystem. Biogeocenosis then includes those aspects of the ecosystem which have an effect on, or are affected by, others, and reciprocity is the key word.

Some other remarks by Morgan and Moss help to confirm a profound change in emphasis which has taken place so gradually in geography. The change must be familiar to all geographers yet it is difficult to date or document. The very crude ideas of determinism early this century usually involved the idea of rock, relief, climate or soil affecting man in the same place. At the same time the concept of the formal region was much more familiar than that of the functional region. Again here was the idea that uniform relief combined with uniform climate and soils to give uniform land use and settlement within a region. In both these instances the co-existence of several phenomena on top of each other or combined on the same site was the basic spatial idea.

In the more recent concept of the functional region, there is variety of relief, soil, land-use, settlement, industry and so on, both within and outside a given region. These phenomena are united not by being piled up on top of one another in the same space, but by functioning and working together as part of an economic and social system. Thus crops, animals, food, raw materials, people, messages are exchanged and moved about by air, water, pipe, wire, road and rail. The basic spatial idea is that these phenomena are spread out horizontally, side by side, and are arranged thus not by chance, but in a logical layout so that they can work together well and make the best use of space. Nowadays the much more refined concepts of determinism must include the idea that a phenomenon here, in this place, can and does affect another there, possibly hundreds of miles away.

Thus Morgan and Moss write of 'causal relationships existing in the complex of heterogeneous phenomena at one place, and the causal connections among phenomena at *different places*'.[16] They write of homogeneous areas or habitats, which are similar to formal regions, inhabited by societies (groups of similar things). Two or more different societies combine to form the community in a functional area, which by name and definition must be similar to a functional region. These communities, of course, like functional regions, are organised and represent small systems or part systems. Again putting the emphasis on reciprocity Morgan and Moss state that 'the purpose is to study the structure and functions of a community within a space or spaces'. But while their paper stresses the interaction, and the involvement of man with the environment to the extent that the division of geography into physical and human is ridiculous, at this point they become concerned only with the specialisation in biogeography.

One difficulty of language must be mentioned here. Morgan and Moss use the word 'relationship' when they refer to several phenomena sited in the same place, and the word 'connection' when they refer to phenomena located side by side in different places. These two words have been used in exactly the opposite way by Brunhes, and earlier in this book. These are basic concepts in geography, and generally accepted conventional terms are needed, but no law can be laid down here in ignorance of what is the more common usage. Finally, the word connection in this book refers to what natural scientists usually mean by the word 'association' in contrast to causal connection.

The parts of the definitions which seem to be most easily misunderstood by the layman and young students are phrases like areal patterns of covariation, site and location, areal differentiation and spatial interaction. Given that the geographer studies interrelated phenomena, important to man, on the earth's surface, then these phrases refer to the way in which the geographer approaches these phenomena. Consideration of this point will help to answer both the questions: What makes a soil geographer and an urban geographer both geographers? and How should a student treat a problem geographically?

Knowing the field, map, and quantitative techniques; realising that certain subject-matter is relevant and necessary to geography, but not unique to it, the geographer examines only certain aspects of phenomena. He examines their distribution throughout the world, the density and localisation of their occurrence, how they are combined and connected with other phenomena in the same place, how they are related to similar and different phenomena in other, distant places. Older, descriptive regional geography stemmed from a question such as 'How is one part of the world different from another?' The question basic to most modern general analytic geography is 'How are phenomena arranged on the earth's surface?' The emphasis is on the bird's eye view, the plan, the map, on geometry and spatial arrangements.

We could imagine minerals and mining as the subject-matter of an historian, a geologist, a mining engineer and a geographer. They approach the common subject-matter in completely different ways. The historian studies the development in time of the mining from the discovery of the mineral until the present day. His interest lies in changing techniques and conditions, possibly in the changing social conditions as well. The geologist studies the chemical and physical properties of the mineral, its association with other minerals, rocks and fossils. He may also go on to study the time and nature of its formation and deposition to give some idea of where such a mineral may be found and where it can be mined profitably. While the mining engineer may well be interested in the history of mining and the nature of the minerals, his basic concern is with how the mineral is extracted now. His study is of the machinery and organisation of the mines, and the processes of mining now.

The author is well aware that some aspects of mining history in the industrial revolution, some mineralogy and geology, some details of mining processes, are taught as aspects of geography in British schools now. But the approach which distinguishes the geographer from these other specialists lies in the particular aspect of mining which he studies. These include the distribution of all such mines throughout a region, country or even the world and their connection *in situ* with the rock, relief, local economic conditions, settlement and labour supply. His interest may lead him to study the typical mining layout and landscape. Further afield, his study must include the location of the mines or mining area in relation to other relevant phenomena, such as the supply of capital and transport, the markets for the minerals, refineries and the industries which use it, supplies of fuel, water, food and so on for the mines.

Such a study would cover all the parts of Hartshorne's most comprehensive definition of geography. The minerals are important to man in making a living; they are necessarily connected with the earth, not an imaginary flat surface. There is the examination of both the interrelationships in place between such things as the rock, minerals and shape of the surface, and the interrelationships horizontally over a distance between such things as mines and factories. Finally, the mining landscape and the industrial system are significant features of the areal differentiation of the earth's surface. In studying a topic such as mining in this way the geographer is concerned with distribution, location, the real combination of phenomena and the actual earth's surface as it is now. He does not need to be concerned with how and why the mining started centuries ago, how the mining has changed in time, the crystal structure or chemical composition of the mineral, nor how the miners use certain methods and machines to get it to the surface.

This last statement may seem to be extreme when so many 'geography' books do include such matters. This is a case in point of the danger of defining geography by what some geographers do. If the best geography consisted of such encyclopaedic knowledge it could not logically exist as a distinctive discipline.

The college student or undergraduate attempting some genuine geographic research for the first time needs much help from his tutor if he is not to be misled by the older kinds of geography which is still common in schools. Sometimes the unaided student thesis becomes an account of processes, whether physical or industrial, but the most common fault is for it to become an

historical narrative. Literature, history, film and television, the
love of a sequential story are so strong in our education that this
temptation must be very hard to resist in geography. Possibly
natural scientists have difficulty in getting students to adapt
themselves to scientific method similar to the geographer's difficulty of thinking in terms of space rather than time.

The inclusion of geomorphology under the heading of geography in schools is one of the major obstacles. When students
fail to tackle the distribution and juxtaposition of landforms and
start to tell the story of how they developed, one sees at once that
the culprit is denudation chronology. It must take a supreme
effort of will-power to shake off the narrative of the erosion cycle
or sequent occupance of the British Isles.

This point brings up the question of whether school geography
should be changed radically to bring it into line with university
geography. If it were changed, the sixth-former would not have
such a traumatic experience on reaching college, and would grasp
the true nature of modern geography much more quickly. But
this would be for the sake of the 2 per cent who will become
future professional geographers[17] and for the sake of the other
98 per cent there is much to be said for school work, perhaps
under another name, to carry on with life and work, regions and
world economic and social problems.

In addition to the narrative tradition, the student suffers from
another great obstacle in trying to adopt the geographical approach in his first thesis. Usually time is so short, and the size of
the area or the subject-matter of the study so limited, that it is very
difficult for him to achieve the scope necessary to reveal the
relationships and integration of phenomena. A worthwhile study
of space relationships usually involves consideration of a whole
country or continent, while any one phenomenon may be involved with others at a place in so complicated a way that a
lengthy study is necessary. There is the danger of a result so superficial that it is as useless as the crude over-simplified determinist
statements of the past. Far from being intellectually satisfying and
worthwhile, it may be downright misleading in its superficiality.

Again one would stress the complexity of both the working of
determinism and the interaction of phenomena on the earth's
surface. One hopes that in addition to everything else the student
is not burdened with having to reconstruct a mythical virgin,
natural environment before studying the effect of this on human

activities. The dangers are manifold. First, this gives a bias to a narrative account, starting with the pre-Cambrian, passing through plant re-colonisation after the ice retreat, through to the discovery of North Sea gas. Consideration of distribution and location never gets a chance. Second, the belief in one way cause and effect is perpetuated. Third, even if this can be overcome, and man's effect on the physical environment is examined, this is only interaction on the site. Thus the chances of an examination of interaction between different places in a horizontal direction are minimised.

Some believe that the main revolution in geography has been not the introduction of quantitative techniques, but the insistence on true geographic method, and the use of inference, classification, systems theory and models. While these are more important they are more difficult to learn, and what the student needs more than instruction in topology, x^2 correlation and regression lines is prolonged guidance in thinking in certain ways relevant to his work and problems. In particular, what many need is greater understanding of the operation of cause and effect, of areal connection both horizontally and vertically. Geographers as a whole would welcome more techniques which help to analyse space, areas, and areal connections.

Just as some definitions embrace all geography, and some just particular parts of it, the purposes of geography vary in their scope and objectives. Among the purposes of individual geographers one may list:

(1) The description of different regions
(2) The understanding of the influence of natural environment on man
(3) Social and economic planning
(4) The understanding of how phenomena combine
(5) The understanding of spatial distributions
(6) The formulation of laws about behaviour in space
(7) The construction of models illustrating arrangements in space.

It was suggested at the end of the last chapter that possibly these, and other, purposes are so diverse that only their approach unites geographers. There is much argument about just what is the unifying feature of geography. For Wrigley it is the subject-matter, for Berry the concepts and processes, and for Lukermann its purpose. In the light of this disagreement among authorities, the present author can state only a personal opinion.

As subject-matter and techniques are property common to many disciplines, these by themselves cannot distinguish geography. The modern geographic approach, which has been examined in some detail, is much more important in uniting the many necessary branches of geography, and in distinguishing geography from other disciplines. The confusion arises because of the existence of many pieces of work which are part geography and part something else. But the purposes listed above, the purposes of individual geographers, seem to be capable of combination into the purpose of geography as a whole.

Borchert[18] writes of 'the basic intellectual problem which has motivated geographic study—the need to see one's self in perspective within the changing patterns of man's resources and activities on the face of the earth'. One might prefer the phrase 'different patterns' to 'changing patterns' but Borchert has touched on a vital point. Each individual is personally involved in geography in a way in which he can never be involved in history or with the objects of study of natural science. We are confined within our own bodies, to our own time, and each to a particular place on this planet.

One may have an intellectual need to understand the history of events before one's own existence, and to understand the nature of physical phenomena. But in addition to a possible intellectual need to see how one's home fits in in relation to the rest of the features on the earth's surface is the fact that the nature of that surface and the organisation of systems on it determines how one makes a living, and how one lives. Chapman writes of the supreme practical importance of our study-problem to the sheer survival of man.[19]

Reading, for example, Alexander's book on the north-eastern USA[20] one is struck by the fact that even highly civilised, sophisticated, technically advanced people such as the Americans are inescapably involved with the earth's surface. Alexander shows that in the state of Maine the people are having a hard time trying to make a living. But the majority of those people have to do it in that place, now. The majority of people in Maine who can't or won't move have to use that landscape, with all the impedimenta and mistakes of the past around them, in order to live. They cannot wipe the slate clean and arrange their particular piece of the earth's surface as they would like it to be, and as it would permit making a living and living in the most convenient and comfortable way.

Obviously in Maine they are living differently from the way they did in the past, arranging themselves on the surface in a more efficient manner, but they still have to use that environment, make the best of it, and overcome its drawbacks. In an example such as Maine we see the facts of 'man's experience in space', mentioned by Lukermann.

There are disciplines such as psychology which deal with man's mind and body on one hand, and physics which deal with the material world quite separately on the other, but together history and geography are the disciplines which deal with man's involvement with the total of time and space. I think, therefore I exist; the universe is. These are the only two definite statements we can make. For each individual there exists 'me' and 'not-me'. *What* the universe is, whether the 'not-me' of people, places, things and time exists independently of us or only in our minds, is a matter of conjecture. But we seem to exist for a certain time, in a particular place. We may have a psychological existence independent of anything else, and natural scientists try to understand the nature of things independent of man, but history and geography deal with aspects of our whole experience where we are involved with whatever exists outside our minds and bodies.

Just as several aims contribute to the whole of geography, so geography is part of man's aim to understand the nature of his existence. If history is the study of such things as man, and human interactions, cause and effect in time, development and change in human activities and institutions, how the past was different from the present, what we are now, and how that came about, then geography is the study of such things as man on the earth's surface, cause and effect in particular places, the interaction of men and things in space, variations in real space from place to place in both concrete objects and human activity, how and why the rest of the world is different from our own place and what we are, here, and how we fit in. When Hartshorne writes of 'significance to man' of the things and arrangements which the geographer studies, one takes this to mean significance in making his living and living in the best possible way. Man's only problems are what to do and how to live. Those two problems have to be solved in terms of the kind of animals which men and women are, and the nature of the surface of the particular planet on which we happen to live. If we were amphibians living on Venus then both our experience and method of solving the problems would be different.

So the geographer studies how man lives in the total environment of this earth's surface. He studies both the degree and kind of dependence on that surface for survival. Because man needs a certain amount of space to live and produce food, because he needs to overcome the obstacle of space in moving himself and goods over that surface, and in building his dwellings upon it, then the geographer is concerned with land use, and how man produces recurring spatial and geometric patterns on that surface. Man has produced more or less logical spatial arrangements of crops, mines, factories, and residences on an originally naïvely given surface of rocks, soil, mountains, plains, trees, grass, rain, drought, heat and cold.

Whatever the original natural conditions may have been like, the combinations of rock, relief, and climate combined in different ways to give different types of environment. Moreover, different tribes and races of men, with different minds, came to inhabit these different places, and their minds, faced with the problem of obtaining food and living on that surface, reacted in different ways. The chain of events of original natural environment affecting millions of different minds in different ways probably is as complex as Martin insists. But it is now just as difficult to separate the man-made and man-altered elements of the environment from the 'natural' as it is to sort out the historical chain of cause and effect.

But the present result is important to the geographer. The original natural variations from one part of the world to another, combined with the different reactions of innumerable groups of men, have resulted in the different appearances of, and arrangements on, the surface of the earth which we see today. In studying man's use of the earth, the geographer cannot study one type of use and one type-pattern of spatial behaviour. Both the way the earth's surface is used and the way man arranges his farms, villages, roads, shopping and residential areas, industrial regions and so on varies from one part of the world to another. Therefore, in addition to the points mentioned above, the geographer studies the variations of life and landscape from one part of the world to another. These result both from the differences in the earth's surface and the different reactions of different minds to that physical environment producing what is now man's total worldly environment. This total environment now includes ideas and buildings, crops and politics, taxes and overpopulation, as well as relief, climate and soils.

Referring, for the last time, to man's experience in space, it seems that geography examines this by seeking answers to five basic questions. What is the particular space of our total environment like? How do we use this space now (not how we have to, nor how we did do in the past)? How do we arrange ourselves, our things and our activities in this particular space? To what extent are we involved with the earth's surface? How could we arrange ourselves in, and make use of, this particular place, in a better way?

1. Berry, B. J. L., 'Approaches to Regional Analysis', *AAAG*, vol. 54, March 1964, p. 2
2. Taylor, G., Ed., *Geography in the Twentieth Century*, Methuen, 1957, p. 106.
3. Brown, R. H., *Historical Geography of the United States*, Harcourt, Brace and World, 1948
4. Darby, H. C., *The Domesday Geography of Eastern England*, CUP, 1952
5. East, W. G., *An Historical Geography of Europe*, Methuen, 1935
6. Beresford, M., *The Lost Villages of England*, Lutterworth, 1954
7. Lewthwaite, G. R., 'Environmentalism and Determinism', *AAAG*, vol. 56, March 1966, p. 3
8. Minshull, R. M., *Regional Geography*, HUL, 1967
9. Eyre, S. R., 'Determinism and the Ecological approach to Geography', *Geography*, vol. XLIX, November 1964
10. Berry, op. cit.
11. Haggett, P., in Chorley, R. J., and Haggett, P., *Frontiers in Geographical Teaching*, Methuen, 1965, p. 374
12. Hodder, B. W., *Economic Development in the Tropics*, Methuen, 1968, p. 22
13. James, P. E., in Cohen, S. B., *Problems and Trends in American Geography*, Basic Books, 1967, p. 11
14. Morgan, W. B., and Moss, R. P., 'Geography and Ecology', *AAAG*, vol. 55, 1965, p. 1
15. Eyre, op. cit., p. 369
16. My italics
17. Henderson, H. C. K., 'Geography's Balance Sheet', *TIBG*, no. 45, 1968, p. 1
18. Borchert, J. R., in Cohen, S. B., *Problems and Trends in American Geography*, 1967, p. 268
19. Chapman, J. D., 'The Status of Geography', *The Canadian Geographer*, vol. X, no. 3, 1966
20. Alexander, L. M., *The Northeastern United States*, Van Nostrand, 1967

BIBLIOGRAPHY

Berry, B. L. J., and Marble, D. F., *Spatial Analysis,* Prentice Hall, 1968

Broek, J. O. M., *Geography: Its Scope and Spirit,* Charles E. Merrill Inc., 1965

Brunhes, J., *Human Geography,* Harrap, 1952

Bunge, W., *Theoretical Geography,* University of Lund, Sweden, Gleerup, 1966

Chorley, R. J., and Haggett, P., *Frontiers in Geographical Teaching,* Methuen, 1965

Chorley, R. J., and Haggett, P., *Models in Geography,* Methuen, 1967

Cohen, S. B., *Problems in American Geography,* Basic Books, 1967

Cole, J. P., and King, C. A. M., *Quantitative Geography,* John Wiley, 1968

de la Blache P. V., *Principles of Human Geography,* Constable, 1926

Dickinson, R. E., *The Makers of Modern Geography,* Routledge & Kegan Paul, 1969

Dickinson, R. E., and Howarth, O. J. R., *The Making of Geography,* Oxford, 1932

Febvre, L., *A Geographical Introduction to History,* Routledge, 1925

Fischer, E., Campbell, R. D., and Miller, E. S., *A Question of Place: The Development of Geographic Thought,* Beatty, Arlington, Virginia, 1969

Freeman, T. W., *A Hundred Years of Geography,* Duckworth, 1961

Gilbert, E. W., *British Pioneers in Geography,* David & Charles, 1972

Gregory, S., *Statistical Methods and the Geographer,* Longmans, 1968

Haggett, P., *Locational Analysis in Human Geography,* Arnold, 1965

Hartshorne, R., 'The Nature of Geography', *AAAG,* Lancaster, Pennsylvania, 1939

Hartshorne, R., *Perspective on the Nature of Geography,* Murray, 1959

Harvey, D., *Explanation in Geography,* Arnold, 1969

Houston, J. M., *A Social Geography of Europe,* Duckworth, 1953

James, P. E., and Jones, C. F. (Eds.), *American Geography: Inventory and Prospect,* Syracuse University Press, New York, 1954

Jones, E., *Human Geography,* Chatto & Windus, 1964

Minshull, R. M., *Regional Geography: Theory and Practice,* HUL, 1967

Perpillou, A. V., *Human Geography,* Longmans, 1966

Sorre, M., *Les Fondements de la Géographie Humaine,* Armand Colin, 1952

Taylor, G., *Geography in the Twentieth Century,* Methuen, 1957

Wooldridge, S. W., and East, W. G., *The Spirit and Purpose of Geography,* HUL, 1951

ARTICLES

Berry, B. L. J., 'Approaches to regional analysis', *AAAG,* vol. 54, March 1964

Chapman, J. D., 'The status of geography', *The Canadian Geographer,* vol. X, no. 3, 1966

Chisholm, M., 'General systems theory and geography', *TIBG,* no. 42, December 1967

Clark, K. T. G., 'Certain underpinnings of our arguments in human geography', *TIBG,* 1950

Claval, P., 'Qu'est-ce que la géographie?', *Geographical Journal,* vol. 133, March 1967

Davies, W. K. D., 'Theory, science and geography', *Tijdschrift voor Economische und Sociale Geografie,* vol. 57, no. 4, July 1966

Eyre, S. R., 'Determinism and the ecological approach to geography', *Geography,* vol. XLIX, pt. 4, 1964

Fisher, C. A., 'Whither Regional Geography?' *Geography,* vol. LV, 1970, p. 373

Floyd, B., 'The Quantitative and Model-Building Revolution in Geography', *Geography,* vol. LV, 1970, p. 40

Golledge, R., and Amadeo, D., 'On laws in geography', *AAAG,* vol. 58, 1968

Harvey, D. W., 'Pattern, process and the scale problem in geographical research', *TIBG,* no. 45, September 1968

Lewthwaite, G. R., 'Environmentalism and determinism: a search for clarification', *AAAG,* vol. 56, 1966

Lukermann, F., 'Geography as a formal intellectual discipline, and the way in which it contributes to human knowledge', *The Canadian Geographer,* vol. VIII, no. 4, 1964

Martin, A. F., 'The necessity for determinism', *TIBG,* no. 17, 1951

Morgan, W. B., and Moss, R. P., 'Geography and ecology', *AAAG,* vol. 55, 1965

Saey, P., 'A new orientation of geography', *Bulletin de la Société Belge d'Etudes Géographiques,* tome 37, no. 1, 1968

Simons, M., 'What is a geographic factor?', *Geography,* vol. 51, November 1960

INDEX